OXFORD MEDICAL PUBLICATIONS

OXFORD TEXTBOOK OF FUNCTIONAL ANATOMY

VOLUME 1

OXFORD TEXTBOOK OF FUNCTIONAL ANATOMY

VOLUME 1
Musculoskeletal system

by PAMELA C. B. MacKINNON
and JOHN F. MORRIS

Department of Human Anatomy, University of Oxford

OXFORD UNIVERSITY PRESS
1986

Oxford University Press, Walton Street, Oxford OX2 6DP

Oxford   New York Toronto
Delhi Bombay Calcutta Madras Karachi
Petaling Jaya   Singapore   Hong Kong   Tokyo
Nairobi Dar es Salaam Cape Town
Melbourne Auckland

and associated companies in
Beirut Berlin Ibadan Nicosia

Oxford is a trade mark of Oxford University Press

Published in the United States
by Oxford University Press, New York

© *Pamela C. B. MacKinnon and John F. Morris, 1986*

British Library Cataloguing in Publication Data
Mackinnon, Pamela C.B.
Oxford textbook of functional anatomy.
Vol. 1: Musculoskeletal system
1. Anatomy, Human
I. Title   II. Morris, John F.
611   QM23.2

ISBN 0–19–261517–3

Library of Congress Cataloging in Publication Data
MacKinnon, Pamela C.B.
Oxford textbook of functional anatomy.
(Oxford medical publications)
Includes index.
Contents: v. 1. Musculoskeletal system.
1. Anatomy, Human—Collected works.   I. Morris, John F.
II. Title.   III. Series. [DNLM: 1. Anatomy.   QS 4 M1580]
QM5.M33   1985   611   85-7099

ISBN 0–19–261517–3 (pbk.: v. 1)

Set in Linotron 202 by
Graphicraft Typesetters Limited HK
Printed in Hong Kong

Foreword

By Charles G. Phillips BSc, MA, DM, FRCP, FRS
*Formerly Dr Lee's Professor of Anatomy in the University of Oxford;
and Emeritus Fellow of Trinity and Hertford Colleges, Oxford*

In my final years as a University teacher-researcher I had the great pleasure of collaborating with Pamela MacKinnon and John Morris in supervising the practical work of first-year undergraduates who were learning living anatomy under the guidance and within a framework of notes written by these two experienced teachers of anatomy — notes which they were improving every year in response to the encouragement of the undergraduates, and which were gradually taking shape as a book that could be found interesting and useful by a wider public.

The General Medical Council's Recommendations of 1957 were liberating in two ways: first, in encouraging Universities to diversify their approaches to medical education instead of compelling them, as hitherto, to conform to a common pattern; and secondly, in designating detailed topographical anatomy as a postgraduate study. A detailed knowledge of topographical anatomy, unless refreshed by daily practical application, can be held in memory only by exceptional individuals; and since a proportion only of medical undergraduates are intending to, or will ever become, surgeons, it is important that medical undergraduates in general should not have their education distorted by spending excessive time and energy on dissection of the whole body. The old idea that every newly qualified doctor should be ready to perform major surgery on any kitchen table has been dead for many years. Aspiring surgeons must nowadays spend two or more post-registration years before being entrusted with independent responsibility, and during those years they must make sure of all the topographical detail they have to know. As for preclinical education, more than lip service has now to be paid to the idea that biomedical science is advancing so rapidly that continuous reorientation will be needed throughout professional life, and that an essential minimal core of *unchanging* fact will have to be carefully defined.

Thus the question arises: what should be the content and emphasis of first-year anatomy? Different schools will — and should — make different approaches to the problem. The course that is being evolved by Drs MacKinnon and Morris and their colleagues is firmly centred on familiarity with the structure and macroscopic functions of the *living* body, investigated as far as possible with eye and hand complemented by modern imaging techniques, clinical and electrical work on the action of muscles, study of prosections made by surgeons-in-training, and careful dissection of limited regions of the body. Above all, the objectives of first-year anatomy are clearly proclaimed in their book. Without definite guidance it is all too easy for undergraduates to suppose that they are expected to memorize the whole contents of the great atlases and encyclopaedias of anatomy. This course is, rather, a training in accurate observation and systematic recording: the observing to continue, and the records to accumulate, throughout professional life. Every first-year anatomist must understand and remember the principles of first aid. Beyond this, only exceptional people can remember all they need to know without ever having to refer to their books and personal records. Such records ought obviously to be consulted by every clinical student who knows he will have to present a case on a ward-round. And every doctor who takes part in an expedition to outlandish places ought obviously to prepare her- or himself to intervene, if necessary, in one of the few life-threatening emergencies that might arise.

Acknowledgements

Any textbook of Anatomy must necessarily reflect, in part, the instruction of our teachers and the enthusiasms of our colleagues over the years. To all of these, and in particular Professors the late Sir Wilfred LeGros Clark FRS, the late G.W. Harris, FRS, Barry Cross, FRS, Charles Phillips, FRS, Ruth Bowden and Joseph Yoffey, our warm and grateful thanks for their expertise and encouragement.

Some of our colleagues, to whom we are indebted, have made important contributions to particular sections of this book: Mr Eric Denman has advised on many surgical points; Professor Charles Phillips FRS initiated the section on the experimental testing of elemental movements; Dr George Gordon initiated the section on the experimental testing of cutaneous sensation; Drs Basil Shepstone and Stephen Golding provided the chapter on Medical Imaging and many of the radiographs; while Dr Daniel Nolan provided other radiographs and obtained some of the arteriograms from Dr Walter Fletcher; Dr D. Linsell provided the ultrasound images. All the artwork has been drawn by Audrey Besterman who so valiantly put up with our comings and goings; many of the photographic illustrations were made by the photographic unit at the Department of Human Anatomy, Oxford, under the direction of Brian Archer; Robert Underwood designed and made the electronic apparatus referred to in Chapter 3. Jane Ballinger was responsible for much of the typing. Many of our colleagues, in particular Lady Margaret Florey and Dr Margaret Matthews, have read and made helpful comments on earlier versions of this book which have been used for teaching in Oxford. And last but not least, since all anatomists are dependent on dissected material, we gratefully acknowledge the careful conservation of prosections by Roger White.

Contents

CHAPTER 1

Introduction

Expectations and use of the book

In writing this book we have deviated from traditional approaches and attempted to emphasize functional and living anatomy and its imaging in medical practice. The book has been designed to enable students of morphology, whether they be medical or dental, physiotherapists or radiographers, nurses or any other student of the biology of Man, to understand the functional design of their own body and that of others.

To dissect an entire cadaver from head to toe can be very instructive and is essential for aspiring surgeons. However, the recent escalation of knowledge in all branches of the preclinical and clinical medical sciences, for which time must be provided in medical teaching programmes, makes this impractical in most courses. Nevertheless, the understanding of how the body is built and how it works is fundamental to the practice of all aspects of Medicine.

This book (Vols. 1, 2, 3) therefore guides its readers through the fundamentals of anatomy using appropriate prosections and a partial dissection of the body. Each seminar has been designed to cover an aspect of the musculo-skeletal system which can be studied in about a 2h period, in a combination we have found suitable. Different courses use dissection and prosections in different proportions. The book is not intended as a comprehensive dissecting manual. Looking at the Contents it might appear that undue emphasis has been placed on the peripheral nerves to the upper limb. However, we anticipate that the upper limb will be studied first. Armed with an understanding of the arrangements in the upper limb, a study of the lower limb is considerably easier since the general principles are similar. This is reflected in the smaller number of seminars devoted to the lower limb and in particular to its nerve supply.

For those who do not have access to prosections and radiographs the book can also be used by referring to the reader's own body and to the illustrations. Liberal reference should be made to embryology, histology, and neuroanatomy and the book should be read in conjunction with physiology and biochemistry texts. It is important that you are able to answer the questions which have been inserted throughout the book; if the foregoing text has been absorbed and understood then they should not prove difficult. Personal notes on your anatomical investigations made in the margins of the book will prove invaluable. At some future time, rapid perusal of a seminar containing additional comments of your own will ensure an equally rapid recall of the basic information. This is a book which, if used properly, should never be discarded.

Living Anatomy is the structure and function of the body in its living state: it is a study of our skeletal system which protects the vital organs and gives attachment to muscles; of muscles and joints which provide for movement between the various skeletal units; of the highly specialized cardiovascular system through which oxygen and nutrients are pumped to individual cells of the body and waste materials are collected for excretion; of the various organs within the head and neck, thorax, and abdomen which enable the body to remain viable by ensuring a homeostatic environment for each individual cell; of the reproductive system which ensures continuity of the species; and of the nervous system which receives and integrates information from both the internal and external environments and which, through our speech, movements, and behaviour, enables us to express our individual character and personality.

The changing form of the body and its relations to function

Always remember that 'the body' is not a single assembly–line product. It develops in the uterus and this development usually produces a 'normal' individual. It develops and grows further during childhood and adolescence to produce the adult form. On occasions, any part of this development may be imperfect to a greater or lesser degree. It will also be obvious to you that variation among normal individuals exists. It is important that you develop a concept of the **range of normality** so that you can judge what is abnormal. For this purpose many illustrations of abnormalities are included in the book. In later adult life, ageing changes lead to senescence. Remember that most bodies donated for examination in the dissecting rooms are those of elderly people.

External differences between male and female are mostly obvious; however, the body of both sexes undergoes many cyclic changes of a circhoral and diurnal nature and that of the female undergoes in addition a monthly cycle. Throughout life, the body responds morphologically to functional demands (e.g. muscle hypertrophy) and to abuses and injuries by repair and healing.

'The body' which you must consider is therefore not the static, usually elderly form which you see

1.1

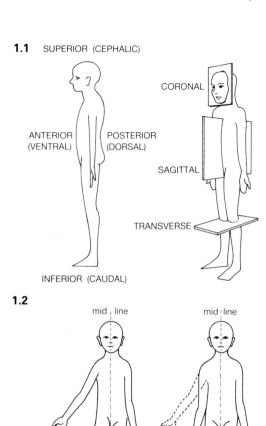

SUPERIOR (CEPHALIC)

CORONAL

ANTERIOR (VENTRAL)   POSTERIOR (DORSAL)

SAGITTAL

TRANSVERSE

INFERIOR (CAUDAL)

1.2

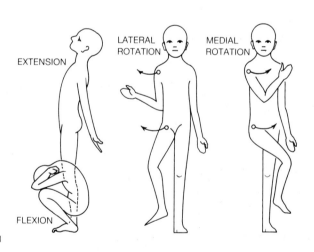

mid line   mid line

ABDUCTION   ADDUCTION

EXTENSION   LATERAL ROTATION   MEDIAL ROTATION

FLEXION

LATERAL FLEXION

PIVOT

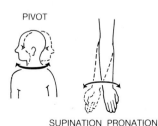

SUPINATION PRONATION

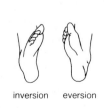

inversion   eversion

on the dissecting room table, but rather a living, dynamic organism, constantly changing and responding to the functional challenges of its environment.

Any region of the body consists of a number of different tissues. The anatomical form that these tissues take in any region is the result of their evolution to fulfil a functional role. The structure of tissues can be considered on two levels. The microscopic structure is the object of study in histology. The macroscopic (naked eye) and radiological structure is the subject of these seminars.

Terms used in accurate anatomical description (1.1, 1.2)

For ease of communication and convenience of description the body is always considered as standing erect, facing ahead, the arms by the sides and the palms of the hands facing forwards with the fingers extended. Place yourself in this position and note that it differs in a number of ways from the normal way in which you stand.

The terms **anterior** (ventral) and **posterior** (dorsal) refer to structures facing the front and back of the body respectively. Structures in the antero–posterior midline are said to be **median**, those close to the midline **medial**, and those further away, **lateral**. Structures above are referred to as **superior** (or cephalic); structures below as **inferior** (or caudal). **Proximal** means nearer to the origin of a structure, **distal** is the opposite. **Superficial** means nearer the skin, **deep** is the opposite.

Anatomical planes

Sagittal   —a vertical plane lying in the antero-posterior plane (longitudinal)

Coronal   —a vertical plane at right angles to the sagittal

Transverse—a horizontal plane at right angles to both coronal and sagittal

Oblique   —any plane that is not coronal, sagittal or transverse.

The movement of a limb away from the midline of the body is termed **abduction** whereas **adduction** is the movement which returns the limb to its original position or towards the midline of the body. A forward or anterior movement of the trunk or limb is usually termed **flexion** while a backward or posterior movement is **extension**. The forearm can be **pronated** so that the palm faces posteriorly, or **supinated** so that the palm faces anteriorly. If the sole of the foot is directed medially it is **inverted**, if it is outwardly directed it is **everted**.

There are certain joints at which **rotation** can occur. If the rotation is towards the midline of the body the joint or body part is said to be **medially rotated**, and if away from the midline, **laterally rotated**; these respective movements can be conveniently demonstrated by swinging the arm with the elbow bent, across the chest and then away from the chest.

If the non–rotated back is kept straight and the head or trunk is bent to either side the movement is referred to as **lateral flexion**.

CHAPTER 2

Systematic thinking about tissues

This section indicates the range of topics concerning a tissue or part of the body that should be considered.

Skin

Consider:

● the **degree of keratinization**: keratinization is protective (compare the sole of the foot and the eyelid) and can be altered by the environment, for instance the calluses caused by manual work.

● the **degree of hairiness** and type of hair: only the palms, soles, eyelids, and penis are hairless. The **density** and **coarseness** of the hair differs from region to region. The **distribution** of body hair is sexually dimorphic, the male pattern being dependent on circulating androgens. Abnormal hair distribution can reflect endocrine imbalance.

● the presence of **dermal ridges**: these function in gripping and probably also in texture recognition as, when the ridged skin is moved over an object of interest, they produce vibrations that are sensed; their pattern forms the finger–print.

● the **sweat glands** and their openings: the openings of **eccrine** sweat glands can be located on the surface of the dermal ridges. Eccrine glands are found over most of the body surface in Man and secrete a colourless watery saline. The number of open active sweat glands on most parts of the body can be increased by exercise and are important in the control of body temperature. Increased emotion or attention however causes activity of sweat glands on the palms and soles (see chapter 3). The **apocrine** sweat glands in the axilla and external genitalia produce sweat with a higher organic content due to cellular breakdown. Bacterial enzymes break down the secretion which results in odour formation.

● the number of **sebaceous glands** and specializations such as areolar glands of the nipple.

● the **skin creases**: at these pronounced lines, which are found especially around joints, the skin is more firmly attached to the underlying tissues. Skin (flexure) creases around joints, however, are different from the fine crease–like lines which appear in the skin of old people. The latter are caused by degeneration of collagen fibres and a reduced attachment of the skin to underlying tissues.

● the **skin cleavage lines** (of Langer): these cannot be seen, but maps are available; in general they run circumferentially round the trunk. They mark the orientation of the fibrous tissue bundles in the skin. Incisions *along* a Langer's line will heal with a fine, barely noticeable scar; those *across* the lines will tend to produce a more 'heaped up' scar (keloid).

● the **blood supply of the skin**: this is derived from local vessels. Capillary loops can be observed if the skin of the nail bed is cleared with a drop of oil and observed with a dissecting microscope. The thermal sensitivity of skin vasculature can readily be demonstrated with hot and iced water. In cold conditions much of the cutaneous blood supply is short–circuited from arterioles to venules by arterio–venous anastomoses.

● the **innervation of the skin**: stimuli to the skin are defined in terms of forms of energy (e.g. mechanical or thermal), their spatial distribution, intensity, and rate of change. It is now clear that some cutaneous sense organs are highly sensitive to mechanical and others to thermal stimuli, and yet others are less specific (see chapter 3 for an experiment).

There are variations in mechanoreceptive spatial discriminatory power in different regions which is usually measured by the two–point discrimination threshold. This is the smallest distance between two simultaneously applied mechanical stimuli that can be perceived as two rather than one.

You will need to know:

● the **local nerve supplying** an area of skin: this is needed both for diagnosis of nerve lesions and application of local anaesthesia.

● the **spinal nerve** which supplies the area of skin: this is necessary for the diagnosis of spinal cord and spinal nerve lesions. The area of skin supplied by a spinal nerve is called a **dermatome**.

Superficial fascia

This is the subcutaneous connective tissue. It varies considerably from place to place.
Consider:

● the degree of **fat** accumulation and its regional variation: compare the thigh and abdomen with eyelids, dorsum of hand, and penis.

● the local accumulation of **fluid** filled sacs: these subcutaneous **bursae** permit the skin to move freely over bony prominences.

● the **fibrous tissue** content. This *a*) determines the attachment of the skin to deeper structures. Compare the palm and the dorsum of the hand. It is prominent at skin creases, and forms the suspensory ligaments of breast, penis, etc. *b*) in some areas fibroelastic sheets are present, such as is found in the superficial fascia of the anterior abdominal wall.

- the presence of **superficial vessels and nerves** passing through the fascia to the skin.

Deep fascia

This is the connective tissue that covers and ensheaths muscles. It is also attached to the bones. Being mainly fibrous in most regions it also contains fat and fluid. Its thickness varies very considerably depending on functional requirements. Over expansile regions such as the pharynx it is very thin, but in other areas such as the leg it forms a non–expansile sleeve which is important in the mechanics of venous return. Where sheets of strong fascia exist they often form sites for muscle attachments. In larger texts you will find that many fascias are named after the muscle that they cover, but only a few of these are important to remember. In certain parts of the body the deep fascia is strongly attached to the skin.

Bones

You should be able to identify any bone and hold it in the position that it occupies in the living body. Consider:

- the position of the **articular surfaces** and the bones with which they articulate.

- **named parts** and **prominences**, especially those that are palpable in the living subject.

- the site of the major **muscle attachments** which may or may not be associated with roughened areas or protrusions of the bone.

- the site of the major **ligament** and **membrane attachments**.

- the **blood supply** and the position of the nutrient arteries.

- the **marrow cavity** content of red or fatty marrow; its extent in both the child and the adult.

- any specializations of the **trabecular** pattern within the bone or thickenings of the cortex which reinforce particular lines of stress in the bone.

- the **ossification** of the bone—and whether it occurs in **membrane** or in **cartilage**.

Primary and secondary centres of ossification produce the diaphysis (shaft) and epiphyses (ends) of long bones respectively. The time of appearance of all these centres need not be committed to memory, but you should acquire a general understanding of the sequence of ossification of the different parts and therefore which is the 'growing end' of long bones. The development of the different centres is useful in determining the age of bones, and in assessing whether the skeletal age of a child matches its chronological age.

Joints

Joints are the articulations between the bones and their form varies widely in relation to the functional requirements of the articulation. The degree of mobility varies widely; some joints permit virtually no movement, in others, only the smallest gliding or angular movements occur, and in yet others, a large range of movement in different planes can occur. Consider the factors that will influence the range of movement. All joints must, however, be **stable**. In joints which have considerable mobility, always consider the specializations which give stability to the mobile structure.

Joint type

Joints are classified as being of different **types** depending on the nature and conformation of the material that separates the bones. Where solid connective tissue (fibrous or cartilaginous tissue) links the bones, the amount of movement possible at the joint will be proportional to the degree of deformation that can be produced in that tissue. This in turn will be determined by the lever force exerted by muscles, on the nature of the tissue and its length in relation to cross–sectional area in the axis of movement.

A. Fibrous joints
The bones of fibrous joints are united by fibrous tissue and there is no appreciable movement between the bony ends unless the fibrous tissue is long in relation to its cross–sectional area, as in the sutures of a child at birth or the interosseous membrane between radius and ulna. Fibrous joints are found in the skull (sutures), between the teeth and jaws (peg and socket), and between bony surfaces connected by an interosseous membrane (e.g. ulna and radius; tibia and fibula) or interosseous ligament (inferior tibio-fibular joint).

B. Cartilaginous joints
These are subdivided into:

i. Primary cartilaginous joints: in which the bones are united by hyaline cartilage. These joints are found between the epiphyses and diaphyses of long bones and also in other places such as between the first rib and the manubrium. Little movement is possible at these joints.

ii. Secondary cartilaginous joints in which articular surfaces covered with hyaline cartilage are united by fibrocartilage. Such joints are all found in the midline of the body: between the bodies of the pubic bones (symphysis pubis), between the bodies of the vertebrae (intervertebral discs); and between the manubrium and sternum. Relatively little movement occurs at these joints, though endocrine–induced changes in the pubic symphysis allow greater movement during late pregnancy and parturition.

C. Synovial joints
In synovial joints the bones, covered with articular cartilage, are separated by a synovial cavity. These are the most common joints in the body. Most allow a considerable amount of movement between the bones though at some virtually no movement occurs. Further classification of synovial joints is based on the shape of the articular surface: **plane**; **hinge**; **pivot**; **condylar**; **ellipsoid**; **saddle**; **ball and socket**. These descriptions

are, of course, only approximations. The shape of the surfaces (together with the ligaments) determines the planes and axes of movement that are possible.

• Consider the **movements** possible at a joint. The nature and range of the movements that can occur will depend on:

a) the shape of the articular surfaces
b) the physical characteristics of the tissues uniting the bones
c) any restrictions by ligaments
d) the mechanical advantage and line of pull of the muscles crossing the joint

The movements can be classified as:

Flexion—Extension
Lateral Flexion
Abduction—Adduction
Medial (internal) rotation—Lateral (external) rotation (of limbs)
Rotation of head, trunk
Pronation—Supination (of the forearm)
Inversion—Eversion (of the foot)
Protraction—Retraction
Elevation—Depression
Opposition of the thumb (and little finger)

• Consider the **stability** of a joint. This depends on:

a) the articular surfaces
b) ligaments
c) muscles

Note that only the muscles provide an active support. If they are paralysed then the ligaments will soon stretch and joint deformity will ensue.

The structure of a joint therefore reflects an evolutionary compromise between mobility and stability in relation to the function of the joint. In any joint note:

• the **articulating surfaces**: the bones that take part, and the shape of their articular surfaces. In some cases the areas of contact in different movements is important.

• the **capsule**: its extent, attachments, strengths and deficiencies. The capsule is usually attached to the margins of the articular surfaces. Note where it deviates from this.

• the **ligaments**: a) Intrinsic ligaments are thickenings of the capsule laid down along particular lines of stress.
b) Accessory ligaments also limit movement of the joint in various directions but are separated from the capsule.

• the **synovial cavity and membrane**: in general the synovial membrane lines all the non–articulating surfaces within a joint. The amount of fluid will depend on the contours of the bones within the cavity but some larger incongruities are taken up by mobile fat pads covered with synovial membrane so that the actual volume of fluid is very low, and is scarcely more than a molecular layer of fluid between the articulating surfaces. On radiographs, the 'space' between the congruent articulating bones is occupied largely by the articular cartilage, which is radiolucent.

• any **bursae**: some are extensions of the synovial membrane out of the joint capsule which provide fluid–filled sacs to give friction–free movement of e.g. tendons over bones. Others related to the joint may not connect with the synovial cavity.

• **intra–articular structures**: **fat pads** covered with synovial membrane help to spread synovial fluid. **Intra–articular discs** of fibrocartilage usually divide the cavity of joints in which movement occurs in two separate axes.

• **the blood supply**: there is usually a good anastomosis of the arteries around joints with a large range of movement, and many of the local arteries give branches to the capsule. This arrangement provides for supply distal to the joint even if the position of the joint tends to reduce the flow in large arteries.

• the **nerve supply**: a) the capsule of joints has a very important **sensory** nerve supply that conveys **mechanoceptive** information to the central nervous system concerning the direction, rate, and acceleration of any movement of the joint, and **pain** fibres. In general, a nerve which supplies a muscle that acts over a given joint will also supply sensory fibres to that joint. More specifically, the part of the capsule which is rendered taut by the contraction of a given muscle or muscle group is usually innervated from the nerve supplying their antagonists. In this way overstretching of part of the capsule is prevented by a reflex. For instance, the nerve to biceps will also supply the anterior part of the capsule of the elbow joint which could be overstretched by excessive contraction of triceps. The capsule has a prominent pain sensing nerve supply that signals excessive movement. An intact nerve supply is thus essential to protect a joint from traumatization due to excess movement.
b) **vasomotor** fibres of the sympathetic nervous system supply arterioles of the synovial membrane.

Muscles and the movements they produce

Many details of muscle topography are not of clinical importance though there are obvious exceptions such as the arrangement of the muscle layers around the inguinal canal in understanding hernias, the arrangement of the tendons of the wrist in trauma surgery, etc. It is generally much more important to understand the muscle groups that produce given movements at a joint and their innervation.

• the actions of muscles: these have been determined in a variety of ways: by performing different movements and feeling which muscles contract; by dissection; by pulling on tendons and inferences from morphology; by electrical stimulation; and by recording by electromyography, cinematography, flashing light photography, and kinesiology.

The muscles acting in any given movement may be classified as:

Prime movers (agonists) which produce the required movement.

Synergists which prevent any unwanted movements which would also be produced if the prime movers acted alone.

Essential fixators which clamp the parts in the position on which the movement is based.

Postural fixators (e.g. of the trunk) which prevent the body being toppled by movements of heavy parts which shift the centre of gravity.

Antagonists which are the muscles which would oppose the prime movers in a particular movement. During the movement they are normally relaxed in proportion to the power of the prime movement.

So-called 'paradoxical' actions may counter the forces of gravity (e.g. biceps contracts when an elbow is extended while lowering a heavy weight).

These combinations of muscles in elemental movements are relatively stereotyped. The stimulation of prime movers, synergists and fixators, and the inhibition of antagonists are determined by what we nowadays call **programmes** laid down in the central nervous system. Some programmes are inherited and enable certain animals to rise onto their limbs and run immediately after birth; others are laid down as the result of learning.

A particular muscle may be a prime mover in one programme, an antagonist in another, a synergist or essential fixator in others.

The number of muscles involved in an elemental movement at a single joint depends:

a) on the mobility of the joint. A hinge joint with its single degree of freedom can be operated by flexor and extensor muscles only. The shoulder joint, with many degrees of freedom, is operated by many muscles in complex, shifting relationships.

b) on the number of joints which the tendon of a prime mover crosses. To operate one of these joints, its actions at the other joints must be annulled by its antagonists at those joints. Those antagonists thus become synergists in the movement at the selected joint. Consider, for example, flexion of the distal finger joints.

Movements more complex than these elemental movements are presumed to be built up by combining the elemental movements. Such complex combinations and sequences cannot be followed by clinical examination alone.

The practical study of the actions of muscles in elemental movements (see p. 9) forms the foundation for more advanced work on how the central nervous system controls movement in health. Furthermore, in cases of acute injury or disease of the spinal segments, roots, plexuses, or motor nerves ('lower motor neurone lesions'), you should be able to determine which of the major muscles are paralyzed. It is no good asking a patient to contract a named muscle, even a professional anatomist cannot do this. You should know what movement to ask the patient to carry out against

resistance in order to prove whether muscle *x*, the belly of which you are feeling, is contracting or not.

Depending on the relationship of the attachment of a muscle to the joint over which it acts, the major component of the force generated by a muscle may act *i*) to produce the movement ('spurt') or *ii*) to maintain the articular surfaces in contact while the movement occurs ('shunt') (**2.1**).

Muscles with a chief role as **prime movers** tend to be attached so that they have a considerable degree of mechanical advantage (e.g. gastrocnemius and soleus); muscles with a primary postural function tend to be shorter and more closely applied to a joint. This is seen particularly well with the muscles of the spine.

Having considered the role of the muscle in movement, note:

Muscle attachments: most skeletal muscles have their primary attachment to separate bones and therefore cross joints. Often they have additional attachments to fibrous tissue the details of which are usually unimportant, and some muscles (especially in the face) are attached to the skin. Visceral, smooth muscle with different properties to those of voluntary muscle usually surround hollow organs.

The **origin** of a muscle is usually described as its more proximal attachment, or the attachment that usually remains fixed during the prime movement that the muscle produces.

The **insertion** is the more distal attachment and usually the part that moves. These terms are really only used for convenience of description since often muscles act in quite different ways in different movements.

An individual muscle fibre can, on contraction, shorten by no more than one third of its length. A muscle therefore cannot originate closer to its insertion than the point at which a ca. 30 per cent change in fibre length would be required by the maximum excursion of the joint. Thus muscles must originate further away from more mobile joints. Consider, for example, the origin of the short scapular muscles from over the medial two thirds (A) of the scapular fossae. Fibres attached at B could not shorten enough (**2.2**).

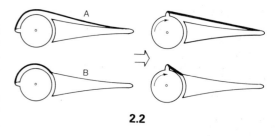

2.2

The **method of attachment**: some muscles appear to be attached directly to bone, though in fact a small amount of fibrous tissue always intervenes. Others are attached via tendons (rounded bundles of fibrous tissue) or aponeuroses (flattened sheets of fibrous tissue). These allow *i*) the bulk of a muscle to be separated from its point of action; *ii*) the pull of a muscle to be concentrated into a very

muscle force

"swing" component

fixed bone

mobile bone

transarticular component

"SPURT"

"SHUNT"

2.1

small area; and *iii*) the line of pull of a muscle to be altered.

The **shape** of a muscle and the **arrangement of its fibres**: two basic equations govern the form of a muscle, both being dependent on the angle at which the fibres are arranged:

a) The degree of shortening is proportional to the length of the muscle fibres

b) The power of a muscle is proportional to the number of muscle fibres (**2.3**).

Muscles with parallel fibres (A) can shorten most, but clearly, for a given volume, the number of fibres can be increased by placing the fibres obliquely to the direction of pull (B); this reduces the length of individual fibres and thus the degree of shortening possible. Muscles are described according to their form. Although the function (and thus the form) of muscles must be understood, memorizing the form of individual muscles is not necessary, though a few examples of different types should be born in mind.

The **nerve supply** to muscles: A knowledge of the **local nerve** of supply to a muscle is needed for the diagnosis of peripheral nerve injuries.

A knowledge of the **spinal segmental nerve** supply (myotome) is needed for the diagnosis of spinal nerve and cord lesions. Always bear in mind that all the muscles producing a given movement of a joint tend to have the same spinal nerve supply (e.g. flexion of the elbow; cervical segments 5 & 6) and that the opposing movement on the joint is often supplied by adjacent spinal segments (e.g. extension of elbow; cervical segments 7 & 8). Posterior primary rami of spinal nerves supply the extensor muscles of the spine; all other muscles are supplied by anterior primary rami. Flexor and extensor muscles (the two basic embryological groups) in any limb are supplied, respectively, by anterior and posterior branches of the anterior primary rami of spinal nerves in the plexuses supplying the limbs.

The **size of the motor units** (number of muscle fibres supplied by one nerve axon) will determine the precision of action possible. For example, muscles producing finger movement have small motor units while those in gluteus maximus are very large.

The **blood supply** to muscles: Although muscles need a good blood supply, the details of the local arteries of supply are rarely of importance. In general, adjacent arteries supply a muscle, and there is a major point of entry of the neurovascular bundle. The two ends of a muscle receive their supply locally.

Arteries

For any artery consider:

● its commencement and parent vessel.

● its **course**, particularly where the vessel's pulsations are palpable (over bone) or the vessel is exposed to injury. Arteries are palpable only at certain points. These are usually where they cross bone; such sites may form useful 'pressure points'

where digital pressure can arrest distal haemorrhage. The course of large vessels can readily be mapped in the living by using an ultrasonic Doppler shift–based probe. Radiologically, vessels can be demonstrated by the injection of radio–opaque material into the circulation at an appropriate point (angiography).

● its **major branches** and **mode of termination**.

● its **area of supply**.

● the **degree of anastomosis** with other major vessels. Some vessels such as the central artery of the retina, are **end arteries**; some are functionally end arteries because of very limited anastomoses; yet others have plentiful anastomoses.

Vessels of the retina can be observed directly in the fundus of the eye using an ophthalmoscope; capillary loops can be visualized in the nail bed. The capillary bed, arteriovenous anastomoses, and other small vessels of the circulatory system cannot be visualized with the naked eye.

The degree of oxygenation and **haemoglobin content** of the blood can best be assessed where the skin is thin at such places as skin creases in the palm, and the conjunctiva of the eyelids.

Veins

For any vein consider:

● its **mode of commencement**, remembering that this is distal.

● its **course**, particularly where the vein can be punctured with a needle for intravenous administration of substances or withdrawal of blood for testing; also areas which are liable to trauma.

● the extent of any **valves** within the veins. There is marked regional variation in the incidence of valves and you should consider the reasons for this.

● the **mode of termination** and **major tributaries**.

● the **area of drainage**.

● the **degree of anastomosis** with other veins, particularly between the **deep** and **superficial** (in relation to deep fascia) veins of the limbs.

Lymphatics

Lymphatic vessels cannot be detected by routine examination of a living subject unless they are inflamed. Also, they are difficult to dissect. They can be demonstrated radiologically after injection of radio–opaque dyes. It is, however, crucially important that you are familiar with the lymphatic drainage of an area since both infection and malignant tumours can spread by this route.

Consider:

● the **lymph vessels** that drain the area: superficial lymphatics tend to run with veins, deep ones tend to run with arteries (as with veins the division between superficial and deep is the deep fascia).

● the **primary receiving lymph nodes** and the extent to which these can be palpated.

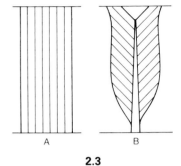

A   B

2.3

● the **route** whereby lymph returns to the blood stream.

● the **degree of anastomosis**: in general there is very considerable anastomosis between the lymphatics serving adjacent areas so that, when lymphatics are blocked by tumour, nodes not normally draining an area may become involved. However, in the leg, there is relatively little anastomosis between the superficial and deep lymphatics.

Nerves

For any nerve that you study consider:

● the **types of fibre** that it contains.

a) **Somatic**—serving the skin and musculoskeletal system

b) **Autonomic**—serving the viscera, glands and blood vessels

The autonomic system is subdivided into **sympathetic** and **parasympathetic** components which differ in their origin, distribution and function. In both systems fibres may be:

a) **motor** (efferent) or *b)* **sensory** (afferent).
In the limbs most nerves contain sympathetic fibres and their somatic component may be motor, sensory or mixed.

● the **origin** of the various fibres:
for somatic nerves the spinal root, and immediate nerve of origin; for autonomic nerves the ganglion or cranial nerve of origin. Note that many sympathetic fibres travel as a plexus around major vessels.

● the **course** of the nerve, particularly where it is palpable or liable to trauma.

● its **major branches**.

● the **muscles supplied** and the effect on movement of trauma to the nerve.

● the **skin supplied** and the area of anaesthesia produced by trauma to the nerve.

● the **organs supplied** (autonomic innervation) and the effects on their function. Remember that in the limbs sympathetic fibres are distributed not only to blood vessels, but also to sweat glands and erector pili muscles.

CHAPTER 3

Experimental examination of tissues

The skin

Sweat glands and dermal ridges

A replica (**3.1**) of the skin which shows the dermal ridges and the number of open sweat gland openings can easily be made by painting the area with a solution of Formvar (4%) in ethylene dichloride, with Sudan black dye (1.5%) and dibutylphthalate (2%) (a plasticizer) added. When dry, the replica can be stripped off with adhesive tape attached to a microscope side, and examined with a low power (×4) microscope objective. The differing distributions of sweat glands in a number of different areas e.g. finger; dorsum of hand; forehead, can be examined. To demonstrate the effect of exercise try making one replica, then exercising by running up and down stairs, finally repeat the replica and compare the number of open sweat glands.

Skin innervation

You can test for the presence of cutaneous sense organs with different responses by exploring the skin *a*) with a fine bristle; *b*) with a warm (40°C) copper object; and *c*) a cold (15–25°C) object. Shaped soldering iron points immersed in warm water or iced water are suitable instruments with which to do this. Repeat your observations on the forehead and on the hand and note spatial variation in the threshold for touch sensation which is related, amongst other things, to the distribution of hair follicles. Cold spots can be very precisely located because the receptors are very superficial, warm spots less so. You will probably find that you can locate separate spots on the palm of the hand but that, on the face, they are too densely packed to be separable. The effects of anaesthesia on two point discrimination can be tested. Locate and mark a cutaneous nerve which supplies an area of forearm skin by exploring along the known course of nerves with a unipolar repetitive electrical stimulus (see *appendix 1*). Mark squares 2 × 2 cm onto the forearm including the area in the centre of the grid supplied by the nerve. Determine and record the variations in two–point threshold over the grid, preferably in both longitudinal and transverse axes because these may differ. Now ask a suitably qualified demonstrator to infiltrate some local anaesthetic around the marked nerve. An area of skin within the grid should become totally anaesthetized within a few minutes. Determine and mark the boundary of the anaesthesia (if a small nerve was infiltrated there may be no completely anaesthetized area). Now determine again the two–point threshold on the grid surrounding the anaesthetic area and also check that the anaesthetic area remains the same. Compare the two sets of data.

Qu. 3A *How do you explain the results?*

Two–point discrimination can be tested using an instrument such as blunted divider points. Test the discriminatory power of various parts, e.g. tongue, finger tip, palm, fore-arm, upper arm, back.

The action of muscles

The actions of muscles can be investigated by different methods. Each method has its advantages and limitations. The results taken together provide more understanding than the results of any one of them taken singly.

Dissection and pulling on tendons (especially in unfixed material). Observations of the effects of pulling on tendons can be combined with inferences from morphology: of the probable effect of shortening the distance between the origin and insertion of a muscle in relation to the possible movements of the joint(s) across which it pulls, and to the pulley–like effect of any ligaments round which its tendon goes. But single muscles do not normally act alone, therefore little is learned by this method of investigation of the natural combinations of muscles which are engaged in producing even the very simplest reflex or voluntary movements.

Electrical stimulation of motor nerves to muscles. In the case of superficially–lying muscles this can be done through the intact skin. The effects of stimulating motor nerves to individual muscles can also be observed when these are exposed at operation. No muscle acts alone, and the actions of single muscles do not correspond with even the most elemental natural movements at a joint. Thus, electrical stimulation leading to an isolated and powerful contraction of the deltoid can cause dislocation of the shoulder! In the natural movement of abduction at the shoulder, the synergic actions of other muscles prevent this.

Clinical examination of the simplest voluntary movements can be carried out at particular joints, to determine which muscles are taking part in them.

Cinematography, flashing-light photography, kinesiology and electromyography. Modern research on human movement involves recording the forces and movements at joints by force transducers, position transducers and accelerometers ('kinesiology'), combined with simultaneous recording of the electrical activity of several of the muscles taking part (electromyography). Movements at several joints, or of the whole body, can be

3.1

analysed by photographing flashing lights attached to the moving parts, or by cinematography, synchronized with electromyography. Cinematography was in fact invented for analysing human and animal movements; its use for entertainment came later.

Stimulation of, and recording from muscles

Stimulation

Motor nerves to muscles can be stimulated so that the individual muscle contracts. This is most conveniently done with an electronic stimulator which delivers 35 rectangular pulses per second, each pulse lasting 0.5 msec. The output voltage is controlled between 5–40 volts (see Appendix Fig 1). A stainless steel anode of large contact area is covered with saturated NaCl–soaked cotton wool and bandaged to some part of the body remote from the part being stimulated. A focal cathode of cotton wool soaked in saturated NaCl solution contained in and just protruding from a perspex tube is applied to the point of stimulation (3.2). The effective 'motor points' are at or near to the point of entry of motor nerves into muscles. Begin with low voltages and, using firm pressure, explore these points, increasing the voltage as necessary to elicit weak contractions. Explore several muscles and note the results.

Recording

As already noted, muscles do not act in isolation, but in a wide variety of different ways in different movements. The electrical activity of a muscle in any movement can be recorded. In clinical practice needle electrodes have to be used for deep muscles, but you can record from superficial muscles through the skin. The subject should recline comfortably on a warm couch and cultivate the art of muscle relaxation. Muscles should first be identified by palpation. Specific movements should then be carried out against resistance offered by the examiners hand.

The areas of skin to which electrodes are to be attached should be cleaned with 70% ethanol and allowed to dry. Attach a large disc earth electrode over the manubrium sterni, and small surface electrodes to either end of the belly of the muscle(s) being investigated. Biceps and triceps are a convenient pair of muscles. These electrodes can then be connected to either an amplifier and chart recorder, or to an amplifier connected to a simple loudspeaker audio device. (see Appendix, Fig. 2)

A possible series of investigations of biceps and triceps—both muscles having parts crossing more than one joint—can reveal a great deal about prime mover action, synergism, essential fixation and action against gravity. We suggest that you investigate their action:

1) in flexion of the elbow against resistance with the forearm both prone and supine;

2) during pronation and supination against resistance;

3) during the gradual squeezing of an object (a dynamometer is ideal) in the hand

4) during flexion and extension; from full extension to full flexion, and back to full extension, with the distal end of the limb weighted in some way—the forearm can be inserted in a tube with a 1kg weight at the end. There are, of course, numerous other possibilities; the essential aim of the experiments is to understand that a muscle has a role in many different movements, rather than a single 'action'.

The circulation

You should investigate on yourself or a colleague all the points at which arterial pulsations can be felt; similarly examine the course of the major superficial veins by lightly restricting venous return to an area. You can also use an ultrasonic Doppler shift–based probe to map larger vessels. If there is one available then choose an area such as the arm, forearm, or palm of the hand and map the course of the arteries and arterial arch.

You should examine capillaries at the two available sites. Clean the skin of the nail bed by application of a small amount of light oil and observe the capillary loops with a binocular dissecting microscope. With the use of an ophthalmoscope view directly the vessels on the surface of the retina in the fundus of the eye.

If you have an inflamed focus such as an infected finger or throat you may be able to feel enlarged, tender and possibly painful lymphatic nodes located respectively in the axilla, or the side of the neck just below the angle of the jaw, which drain these two areas.

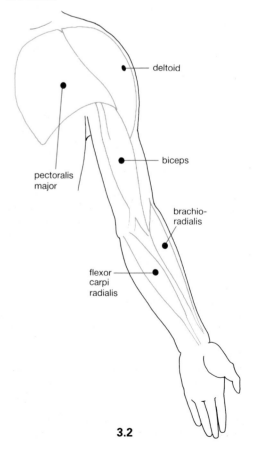

deltoid

biceps

pectoralis major

brachio-radialis

flexor carpi radialis

3.2

CHAPTER 4

Medical imaging

One of the most important adjuncts to physical examination in the study of the anatomy of living subjects is **medical imaging**. Since the turn of the century images (**radiographs**) obtained with x–rays have been the standard method of visualizing many areas of the body. Images can now be obtained using sound waves (**ultrasound imaging**), or by using radiation emitted from substances which have been administered to the patient (**nuclear imaging**). Recently, computerized techniques have been developed which display a cross–section 'slice' of the body. The first of these, **computed tomography** (CT), uses conventional x–rays and its principles have been extended to nuclear medicine to produce **emission computer assisted tomography** (ECAT), and to the **nuclear magnetic resonance** (NMR) effect to produce an image based on magnetic fields.

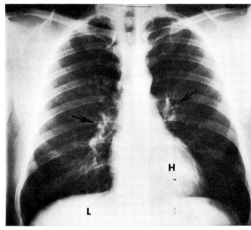

4.1

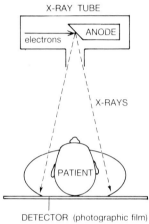

4.2

Radiography

In November 1895 when Wilhelm Roentgen was working on electron beams produced in a Crook's tube, he noticed light being emitted from material on a nearby bench. He realized that some form of emission other than electrons was passing through the tube and causing the material to fluoresce. Placing his hand between the tube and the fluorescing material he saw a shadow of his bones and realized the implications of his observation. By the end of the year he had published an important paper on the newly discovered 'x–rays' which rapidly led to the setting up of primitive x–ray departments in many hospitals.

X–rays are part of the electromagnetic radiation spectrum and can be produced by bombarding a tungsten anode with electrons using high voltages. When the electrons strike the anode their kinetic energy is converted to heat and radiation, including x–rays. In medical radiography the tungsten anode is suspended over the patient so that the beam of x–rays passes through the body; the emerging radiation is then picked up by a detector which is usually photographic film (**4.2**). Since photographic film is sensitive to x–rays, the film will be exposed to a degree that depends on how much of the beam has passed through the patient and how much has been absorbed by the different tissues of the body. Air is radiolucent, bone and metal radio–opaque. In the chest, good contrast is provided by the bone, soft tissues and air–containing lungs, and a clear image of most major structures such as heart (H), liver (L), and pulmonary vessels (arrows) can be obtained (**4.1**). In the abdomen, however, most of the organs are of

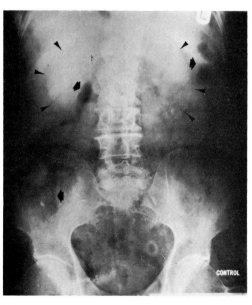

4.3

similar density with respect to x–rays and, although gas in the bowel (◗) and bone can be distinguished, there is very little extra information to be obtained from soft tissue structures although the outlines of the kidneys are just visible (▶) (**4.3**).

Fluoroscopy ('Screening')

If a fluorescent screen is substituted for photographic film a direct image is produced. This enables movements of organs to be studied. It is customary nowadays to view the image on a television screen; an amplifier system or 'image

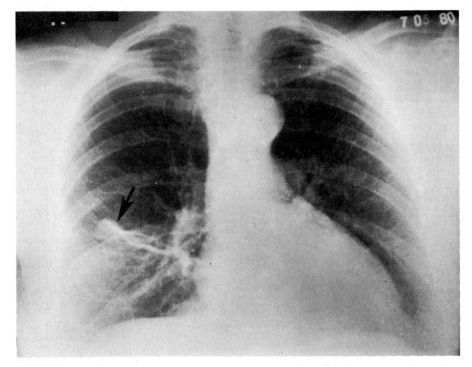

4.4

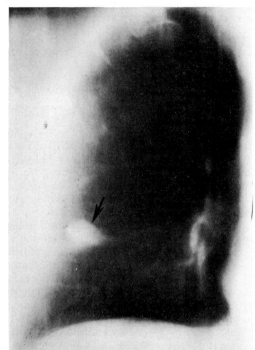

4.5

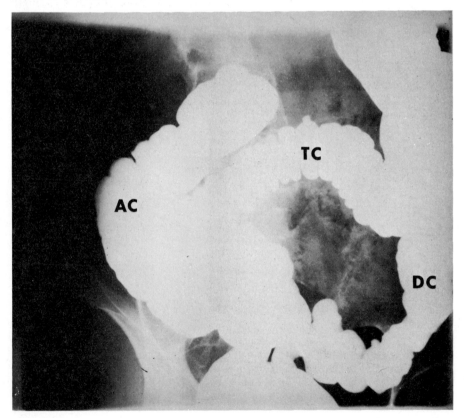

4.6

Tomography

Can be used to overcome the superimposition effects present in the conventional radiographic image. The x–ray tube and the film move around the patient in a constant relationship to the plane of interest which therefore appears as a sharp image; the overlying and underlying areas move relative to the tube and film so that their image is blurred. The effect of tomography is therefore to produce a clear image of one plane only. It is widely used in the investigation of the chest, kidneys and skeleton. Examine the conventional radiograph (**4.4**) which shows an opacity in the right lung. Note that in the tomogram (**4.5**) the plane of focus is such that only the nodule is clearly seen.

Contrast media

It is often not possible to distinguish many organs by conventional radiography, especially in the abdomen. This problem can be overcome by the use of contrast media which are usually either pastes of inorganic barium salts for rectal or oral ingestion, or organic substances containing iodine for intravenous administration. To demonstrate the oesophagus and stomach, a suspension of barium sulphate is given to a patient by mouth. The colon can be similarly studied if barium is introduced through the rectum; note that it has filled the descending colon (DC) transverse colon (TC) and ascending colon (AC) (**4.6**). More information is provided about the lining of the intestine if a small quantity of suspension is used and the gut is then distended with air, giving a 'double contrast' image. In **4.7** barium fills the upper part of the stomach of the recumbent patient, while the mixture of gas and barium allows detail of the lining mucosa to be seen.

intensifier' is usually employed which ensures a good picture with a reduction in the dose of radiation. The fluoroscopic image can then be recorded by photography or video, or by a digital computer which stores the information received (**digital radiography**).

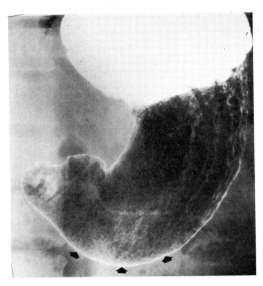

4.7

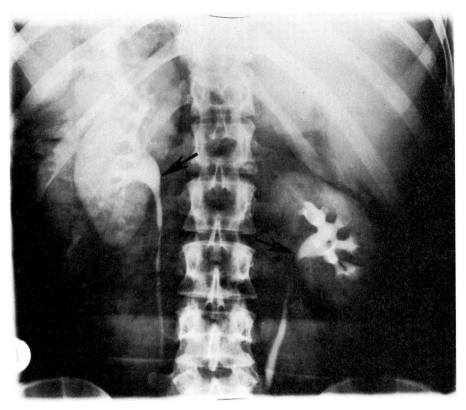

4.8

Iodine contrast media have many applications; they can be injected intravenously to be excreted by the kidneys, (**intravenous urogram**) (**4.8**). The contrast medium is concentrated by the renal parenchyma, making the kidney more obvious than in **4.3** and is excreted into the renal collecting system, renal pelvis, and ureter (arrows). Injections of contrast medium can also be made into a joint (**arthrogram**), into the bronchial tree (**bronchogram**) or around the spinal cord (**myelogram**).

It is also possible to study blood vessels by inserting a fine tube (catheter) into an accessible vessel, for example the femoral artery, and passing its tip into the desired vessel, where contrast media can be injected (**4.9**). The catheter (arrow) has entered the left renal artery (arrow head) and contrast medium has outlined small vessels within the kidney. This procedure, called **angiography**, can be applied to vessels of the gut or head and neck, and even the chambers of the heart.

Ultrasound imaging

Sound waves travelling in a medium are partly reflected when they hit another medium of different consistency. This produces an echo and the time taken for the echo to reach the source of the sound indicates the distance from the reflecting surface. Ultrasound imaging detects and analyses these echos. Ian Donald, Professor of Obstetrics in Glasgow, realized the potential of this technique and developed it to study the pregnant uterus and the fetus.

Ultrasound consists of high frequency sound waves that cannot be detected by the human ear. When ultrasound travels in human tissue it undergoes partial reflection at tissue boundaries; thus a proportion of the sound waves return as an echo while the rest continue (**4.10**). Bone almost totally absorbs the sound and therefore no signal can be obtained from bone or the structures beyond. Neither can a signal be obtained from gas–containing viscera. These two facts are responsible for significant limitations to the use of this technique.

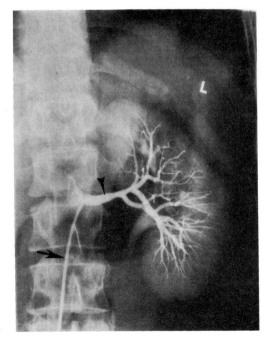

4.9

4.10

However it has the great advantage that there are no damaging effects at the sound energies used and it can therefore be used to monitor the developing fetus. In **4.11** the cranium (C), trunk (T), face (arrow), limbs (arrow head) and placenta (P) of a 15 week fetus can be distinguished. The images are sectional in that they represent a 'slice' of the body

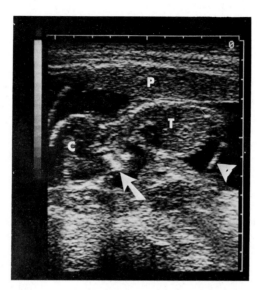

4.11

in the plane of the ultrasound beam. In addition to its use in obstetrics, ultrasound is used in the investigation of the kidney, liver, and biliary system. Recently machines have been developed which have a rapidly repeated scanning action, enabling movement of various organs such as heart valves to be studied ('real–time' ultrasound).

Nuclear imaging

Nuclear medicine may be broadly defined as the use of radioactive isotopes ('radionuclides') in the diagnosis and treatment of disease. Interest to the student of anatomy arises from that branch of nuclear medicine called nuclear imaging or 'scintigraphy', whereby a 'map' of the uptake of radionuclide in a given organ system or pathological lesion can be produced by means of an instrument called a gamma camera. Analysis of such an image (which depends not only on morphology, but also on function) is then of use to the diagnostician in elucidating the nature of the patient's problem.

Historically, the subject began in 1896 with the discovery of radioactivity by Henri Becquerel, followed by the development of the radionuclide tracer method by Georg von Hevesy, the discovery of artificial radionuclides by Irene Curie and her husband, Frederic Joliot, and the invention of the gamma camera by Hal Anger.

The basis of the gamma camera is a large flat crystal of sodium iodide which converts into light rays the gamma rays emitted from radionuclides which have been injected into the patient. These light rays strike a photosensitive surface and cause the emission of electrons. The latter are amplified by a photomultiplier and eventually electrical pulses are formed. These electrical pulses can be used to produce an image on a television screen or be used as input to a computer system (**4.12**).

A number of radionuclides are used in a form described as 'radiopharmaceuticals'. The principal radionuclide used is Technetium–99m (which is very safe for the patient in terms of absorbed radiation dose, and has a convenient half–life of 6 hours), but others are also available e.g. Iodine–123, Thallium–201, Gallium–67, Indium–111. These radionuclides are coupled to various compounds designed to deposit them in the organ of interest. For example, Technetium–99m coupled to sulphur colloid will be phagocytosed by macrophages and so be deposited in the reticuloendothelial system. Technetium–99m phosphates will deposit in bones. In **4.13** note the focal areas of increased isotope uptake in the spine and pelvis

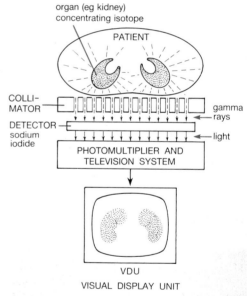

4.12

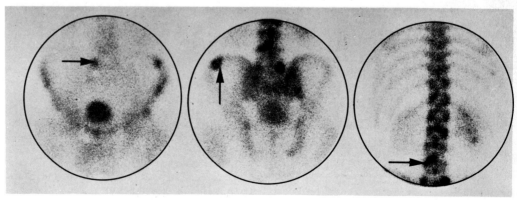

4.13

which are due to tumour deposits from breast cancer (arrows). Iodine–123 as sodium iodide will be taken up by the thyroid. The normal uptake of such a radiopharmaceutical in an organ must be well known to the radiologist, so that any deviation from the normal pattern is rapidly detected.

Cross–sectional images (emission computed tomography) can also be produced using methods very similar to those for emission computed tomography.

Cross-sectional imaging

Computed tomography

In the early 1970s Godfrey Hounsfield, of EMI Ltd, discovered a way of producing very clear cross–sectional images of the body using x–rays. This was hailed as the most significant advance in radiography since Roentgen's discovery and he, like Roentgen, was awarded a Nobel Prize.

Computer tomography (CT) is similar to conventional radiography in that a beam of x–rays is passed through the body and is measured on emerging. The differences between CT and conventional x–ray methods is that a very narrow beam is used and an array of highly sensitive photo-electric cells is substituted for photographic film. The beam is rotated around the patient and density measurements are made from many different angles. The data is analysed by a computer and the whole image is displayed on a TV monitor (**4.14**), dense structures such as bone conventionally being shown at the white end of the spectrum.

CT images are usually true axial sections although coronal sections can be taken of the head by positioning the patient appropriately. However, computer programmes are available whereby a sagittal or coronal image can be built up from data derived from successive axial images. CT produces extremely clear and finely detailed cross–sectional radiographs without any superimposition of surrounding structures. Thus it is possible to see organs or small differences in density caused by disease which are almost impossible to demonstrate by conventional radiography. **4.15** is a CT

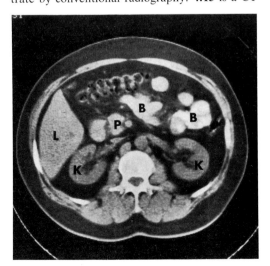

4.15

scan of the abdomen in which the kidneys (K), liver (L), pancreas (P), and small bowel (B) are all clearly seen because they are outlined by body fat.

Emission computer assisted tomography (ECAT)

This is based on the same principle as transmission computed tomography described above, except that it depends on gamma rays emitted from a radionuclide concentrated in an organ rather than on a transmitted x–ray beam (**4.16**). In **4.17** technetium–labelled methylene diphosphonate has

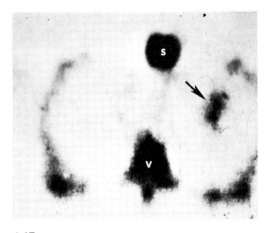

4.17

been taken up in the skeleton of the chest in the same way as in the examination in **4.13**. This however is a cross–sectional image, so that you see the nuclide activity in the vertebral body posteriorly (V) and the sternum anteriorly (S). The uptake within the left side of the thorax (arrow) is in an area of muscle infarction in the heart. This would not have been seen on a conventional scintigram as it would have been obscured by activity in the overlying ribs.

Nuclear magnetic resonance imaging (NMR)

NMR is a new diagnostic technique based on radio signals emitted from resonating atoms within the body. The phenomenon of nuclear magnetic resonance has been known for some time and was developed by Sir Rex Richards, in Oxford, as a means of biochemical estimation of various substances. In the late 1960s this was extended to the analysis of human tissue and the first NMR image was produced by Paul Lauterbur of New York State University.

The NMR effect depends upon the fact that atoms carry a charge and can therefore be regarded as small magnets. When a large magnetic field is applied across a body there is a tendency for the nuclei to line up with this field. Nuclei spin and, rather like spinning tops, can be displaced from their axis by a field applied at an angle to the main

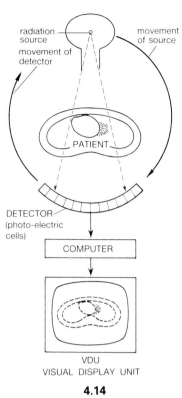

4.14

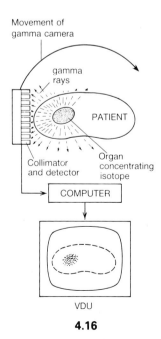

4.16

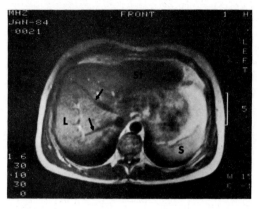

4.18

one. If the second field is of an appropriate frequency the atoms can be made to resonate. Since the atomic nuclei carry a charge, any movement sets up radio signals which can be detected outside the body. The interest of NMR centres on the fact that the behaviour of the resonating nuclei depends upon the atoms surrounding them, in other words their chemical environment. Thus the radio signals measured outside the patient can reflect the biochemistry of the area under examination.

The NMR scanner looks rather like a CT machine in that the patient lies on a couch and enters a gantry which in this case is a very large electromagnet. Sections are taken through the area being examined and the resulting NMR images are superficially similar to CT. However, CT is based on the **density of tissues** to x–rays whereas NMR images reflect some aspects of **the biochemical composition of the tissues.** In **4.18** the section through the abdomen shows liver (L), spleen (S), stomach (St) and hepatic veins and ducts (arrows).

CHAPTER 5

Development of the limbs

Limb buds first appear at about four weeks gestation as small lateral outgrowths of the trunk; the upper limb at the lower cervical level, the lower limb opposite lumbar and upper sacral segments. Each bud comprises a core of mesenchyme covered with ectoderm. At its apex the ectoderm forms a ridge which determines a pre– and post–axial border and dorsal and ventral surfaces. The mesodermal core gives rise to the skeleton, vasculature and dermis of the limbs; muscles are derived by migration of somite mesoderm while the ectoderm gives rise to the epidermis and its derivatives.

As the mesenchymal core grows it adds on at its tip giving rise to progressively more distal components of the limb. The mesenchyme condenses to form the primordia of the skeleton, which will later chondrify and then ossify. Mesoderm from the somites migrates to surround the skeletal primordia and produce the muscles of the limb while, at the same time, ventral rami of adjacent spinal nerves (5th cervical to 1st thoracic for the upper limb; 2nd lumbar to 4th sacral for the lower limb) invade the limb bud. The nerves acquire their motor distribution by invading the nearest unoccupied muscular territory available to them. The more caudal nerve roots grow into the limb slightly later so that the basic pattern of motor innervation is that progressively more distal muscles within the limb are innervated by progressively more caudal segments of the spinal cord. Each ventral ramus forms an anterior division that basically supplies flexor muscles which develop ventrally within the limb, and a posterior division that supplies the extensor (dorsal) muscles. Some muscles such as latissimus dorsi which have extensive attachments to the trunk migrate for considerable distances to gain attachment to the skeleton of the limb.

As the mesenchymal core enlarges, the ectoderm to cover it appears to be progressively derived from the ectoderm of the body wall (**5.1**).

For this reason the skin of the most distal part of the limb is innervated by the middle of the group of nerve roots that supply the limb; more cranial and more caudal roots innervate respectively the preaxial and post axial borders of the limb. In the midline between these two borders, the territory supplied by some roots is so attenuated that non-adjacent nerve roots effectively supply contiguous territories of ectoderm, producing 'axial lines'.

Early on, the upper and lower limb buds develop in a very similar manner. The site of flexion creases of elbow, wrist, knee and ankle become apparent at a stage when the hand and foot are only flattened terminal expansions of the limb buds. Within these expansions the mesenchyme condenses to form the pattern of the digits. Further development occurs by a combination of cellular proliferation and cell death. The latter is very important in shaping areas such as the axilla, in the cavitation of joints and in the separation of the fingers. Failure to form sufficient tissue can lead to conditions such as the very maldeveloped limbs associated with maternal ingestion of Thalidomide; failure of cell death can explain the appearance of syndactyly in which two or more fingers are fused. One or more extra digits may also develop.

The mechanisms which lead to this complex pattern formation are only poorly understood. Current theories are based on the idea that cartilage rudiments condense out of the uniform mesenchymal core using cell properties such as aggregation, with neighbouring centres of aggregation competing. Successively distal elements are laid down in a branching pattern based on more proximal elements which have already been formed so that there is one proximal limb bone—the humerus or femur—but five digits.

By the end of the 6th week of gestation the long axis of each limb is roughly at right angles to the trunk, with the elbow and knee prominences pointing laterally. Thereafter the upper and lower

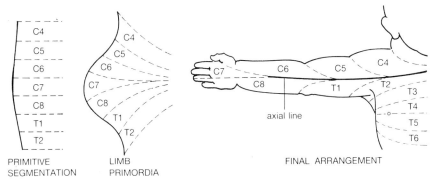

PRIMITIVE
SEGMENTATION

LIMB
PRIMORDIA

FINAL ARRANGEMENT

axial line

C4
C5
C6
C7
C8
T1
T2

C4
C5
C6
C7
C8
T1
T2

C7
C6
C5
C4
C8
T1
T2
T3
T4
T5
T6

limbs undergo different patterns of differential growth. These cause the lower limb to rotate medially such that the knee points cranially and the great toe comes to lie medially; and the upper limb to rotate laterally such that the elbow points caudally and the thumb comes to lie laterally.

In each limb a primary axial artery develops and contributes to the primitive capillary plexus. This main trunk will form the definitive axillary and brachial arteries and, more distally, the anterior interosseous artery and deep palmar arch. In the lower limb the primary axial artery arises from the umbilical artery and follows the developing sciatic nerve. The external iliac artery and its continuation, the femoral artery, develop separately from the common iliac artery and provide a new channel to the lower limb which anastomoses with and eventually takes over the supply of the great majority of the limb.

Veins and lymphatics develop in the limb to form both a deep drainage system which lies alongside the major arteries and a more superficial system in the subcutaneous tissue which eventually drains into the deep vessels.

CHAPTER 6

The upper limb
Introduction

About 75 million years ago when primates appeared in the fossil record, one of the major adaptations to evolve was the increasing dominance of the fore-limb for purposes of climbing, feeding and moving. Joints of the forelimb became more mobile, distal forelimb bones began to rotate about each other, and the thumb was set apart from the rest of the five digits for ease of grasp. With the emergence of the hominids some 72 million years later bipedalism became increasingly successful and with it, the 'emancipation of the forelimb'. Subsequent selection pressures led to the evolution of Homo sapiens with an upper limb set well away from the trunk by a long collar bone, and with the ability both to rotate the hand on the arm through 180° and also to oppose the thumb and forefinger.

These evolutionary adaptations have resulted in an upper limb which is the principle means whereby Man interacts mechanically with his environment. The hand is specialized for grasping items and for moving them with various degrees of power and precision. The ridges which make up the fingerprint, and the firm anchorage of palmar skin both aid such mechanical interaction. The hand can also be used to push or strike objects as can the forearm, arm and shoulder. These more proximal parts of the limb, which contain relatively coarse muscle groups in comparison to the small muscles responsible for exquisitely precise finger movements, can therefore function both in their own right, and to steer and position the hand. These developments in mechanical function could not have taken place in isolation and without parallel changes in control by the central and peripheral nervous systems.

The hand is also the principal means by which Man actively explores his environment. The fingers are therefore particularly richly supplied with sensory receptors which convey mechanical, thermal and painful stimuli to the central nervous system for analysis and evaluation. Movement of the fingers and hand around an object enables us to identify complex objects without looking at them, and the sensory system also provides essential feedback in the control of fine movements.

It is important, therefore, that while studying the upper limb, its role in communication via sensory information and gesturing, in balance, in carrying, in making a power grip, and in delicate manipulative skills, should all be carefully considered.

Seminar 1

Bones of the upper limb

The **aim** of this first seminar is to study the skeletal framework of the upper limb with respect to the living body and to the isolated bones and their radiographic appearances. It is essential therefore to have a partner clad in a sports vest and skirt/trousers; an articulated skeleton and individual bones of the upper limb; radiographs of the upper limb; and a practical notebook for recording your findings by words and drawings.

A. Living anatomy and the bony skeleton

Identify on your partner, on the skeleton, and on the radiographs the bony points listed below (**6.1.1, 6.1.2**):

1. Clavicle

- sternal end
- acromial end
  the sigmoid shape of the bone

The clavicle is the first bone in the body to ossify and it does so by ossification of a membranous model of the bone. All other long bones are formed by ossification of a cartilaginous model. The clavicle is also the only 'long' bone which does not possess a medullary cavity.

The clavicle acts as a strut holding the upper limb, to which it is attached, away from the trunk. Furthermore, it transmits forces from the upper limb to the axial skeleton.

Pick up a clavicle and note the areas of roughness on the under surface of the bone.

Qu. 1A *What forces might have been responsible for these bony protuberances during the course of its development?*

Qu. 1B *Orientate the isolated clavicle and place it as close as possible in its correct position on your partner. How can you tell which is a right or left clavicle?*

2. Scapula

- coracoid process
- acromial process
- spine of the scapula
- supraspinous fossa
- infraspinous fossa
- subscapular fossa
- lateral and vertebral borders
- inferior angle
- glenoid (articular) fossa

Place a scapula as close as possible to its correct position against your partner. Now repeat this with your arm raised above your head and note the change in its position.

3. Humerus

- anatomical head, neck, and shaft of the bone
- surgical neck
- greater tuberosity
- lesser tuberosity
- intertubercular (bicipital) groove
- deltoid tuberosity
- lateral epicondyle
- medial epicondyle
- trochlea
- capitulum
- olecranon fossa
- radial and coronoid fossae
- radial (spiral) groove

4. Ulna

- shaft of the bone
- coronoid process
- olecranon process
- trochlear notch
- facet for articulation with head of radius
- lower end (head) of ulna with its styloid process

5. Radius

- head (proximal end), neck and shaft of the bone
- radial tuberosity (also known as bicipital tuberosity)
- dorsal tubercle on the distal end of the radius (Lister's tubercle)
- lower end of radius with its styloid process and ulnar notch

With the upper limb held straight by the side and the palm facing forwards you will note that the forearm is angled laterally with respect to the arm; this is known as the 'carrying angle'.

6. Carpus

The carpus as a whole is arched transversely due to the shape and articulation of the carpal bones.

- scaphoid, its tubercle, waist, and proximal pole
- lunate
- triquetral
- pisiform (a sesamoid bone)
  } proximal row

- trapezium, its groove and ridge
- trapezoid
- capitate
- hamate, and its hook
  } distal row

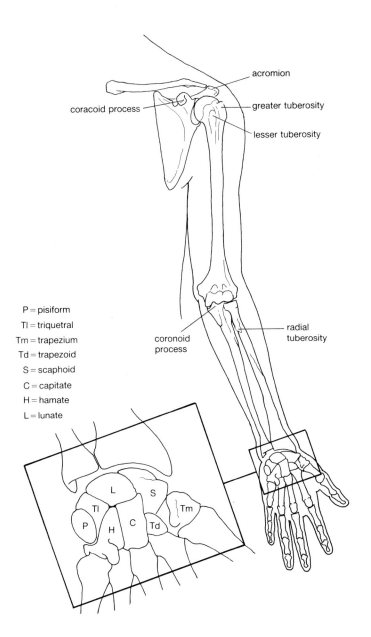

P = pisiform
Tl = triquetral
Tm = trapezium
Td = trapezoid
S = scaphoid
C = capitate
H = hamate
L = lunate

6.1.1

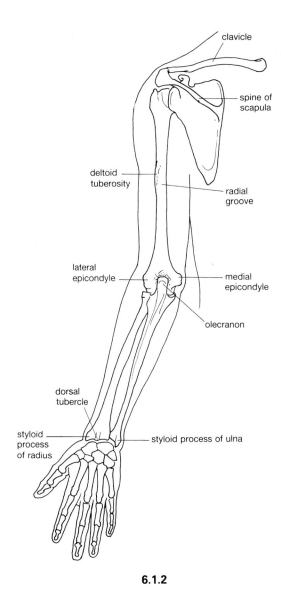

6.1.2

Qu. 1C *Which carpal bone would transmit the greatest force if a person fell onto the outstretched hand?*

7. Metacarpus

Each of the 5 metacarpal bones has a
● head (knuckle)
● shaft
● base (that of the third is particularly prominent dorsally)

8. Phalanges: proximal, middle, and distal

● heads
● shafts
● bases

B. Radiography

When radiographs are examined the following three points should be routinely checked.

i) The outline of the bone, which includes the various prominences already mentioned, is important since the diagnosis of an abnormal appearance depends on a clear knowledge of that which is normal.

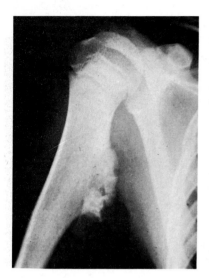

6.1.3

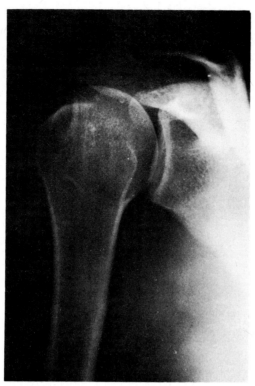

6.1.4

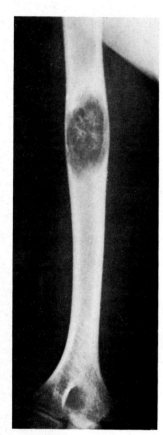

6.1.5

Qu. 1D *Compare the radiographs (6.1.3, 6.1.4) and comment on their appearance; which is abnormal and why?*

ii) The consistency of the bone, note the appearance of the hard outer cortical bone and the less opaque inner cancellous bone, remembering that their relative thickness differs in different bones and in different parts of the same bone. Sometimes the fine bony lines (trabeculae) in the cancellous areas have a characteristic pattern.

Qu. 1E *What abnormality can be seen in 6.1.5?*

iii) The relationship between the ends of bones; where they meet to form joints (this aspect will be dealt with in later seminars).

C. The development of bones

Bones are formed ultimately from the mesoderm of the embryo. Most develop from a cartilaginous model which then ossifies. The clavicle and bones of the skull (except the base) ossify in models composed of condensed mesenchyme (ossification in membrane). Osteogenesis starts in a centre of ossification which frequently lies in a central position in a bone and spreads through the soft tissue model until the whole bone is ossified.

Many bones, particularly smaller bones (such as those of the carpus and tarsus, auditory ossicles, etc.) develop from a single centre. These centres appear over a wide period ranging from the 8th week of intrauterine life to the 10th postnatal year. Many other bones (in particular the long bones and girdle bones of the limbs) are ossified from multiple centres. **The primary centre** usually appears near the middle of the bone (in long bones, the middle of the shaft) early in development, from the 7th (clavicle)–16th week in utero. From this primary centre ossification proceeds towards the ends or periphery of the bone.

Secondary centres appear at the ends (periphery) of the bone from a time just before birth (the centre for the lower end of the femur) onwards to late teenage. This is the mode of ossification of the long bones, metacarpals, metatarsals and phalanges of the limbs, the ribs and the clavicle. Between the ossifying shaft (**diaphysis**) and ends of the bone (**epiphyses**), lie plates of cartilage (**epiphyseal plates**) and at these sites, growth continues. Growth is usually more marked at one end of a long bone (the 'growing end') than the other. Growth gradually ceases towards the end of puberty and the epiphyseal plates become calcified so that the shaft fuses with the epiphysis. Unless one is aware of this pattern of ossification it is possible to mistake an epiphyseal plate for a fracture. Fusion of the epiphyses to the shafts is usually complete by 18–21 years.

It must be emphasized that during the entire growth period, the anlage and the developing bone are being continually remodelled 'keeping pace with' the growth. Varying tensions on bone, such as will be exerted by the insertions of muscles, tendons or ligaments, will alter the contour in that area e.g. the deltoid tuberosity. Abnormal tension or pressure can lead to gross deformity.

Knowledge of the times of appearance, rates of growth, and times of fusion of the secondary centres is often necessary in clinical practice, for instance in assessing the skeletal age of a child in comparison with its chronological age, or in forensic medicine, but these can be looked up at the appropriate time. A radiographer will often radiograph both the normal and the abnormal limb in the same position on one film to assess the symmetry. Examine the radiographs of the upper limb and note the secondary centres of ossification, especially those of the elbow region; fractures of the elbow in children are relatively common.

Identify the bones and epiphyses in 6.1.6–6.1.12. Note especially the scaphoid and the lunate in 6.1.12 and the overlapping articulation of the distal row of carpal bones with the bases of the metacarpals; the overlapping surfaces frequently simulate fractures unless they are carefully delineated.

Qu. 1F *Note the secondary centres of ossification in the fingers. The first metacarpal (of the thumb) might be referred to as a phalanx; why?*

Requirements:
    An articulated skeleton
    Separate bones of the shoulder, arm, forearm, and hand
    Radiographs of the upper limb

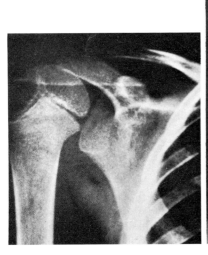

6.1.6

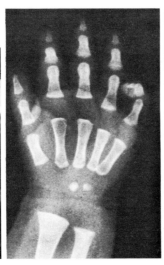

6.1.9

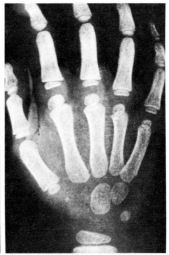

6.1.10

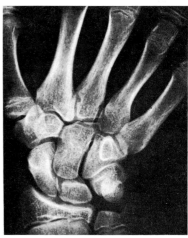

6.1.11

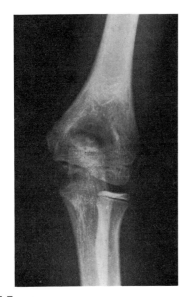

6.1.7

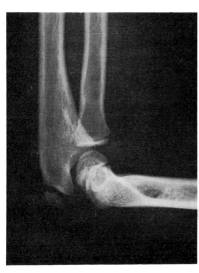

6.1.8

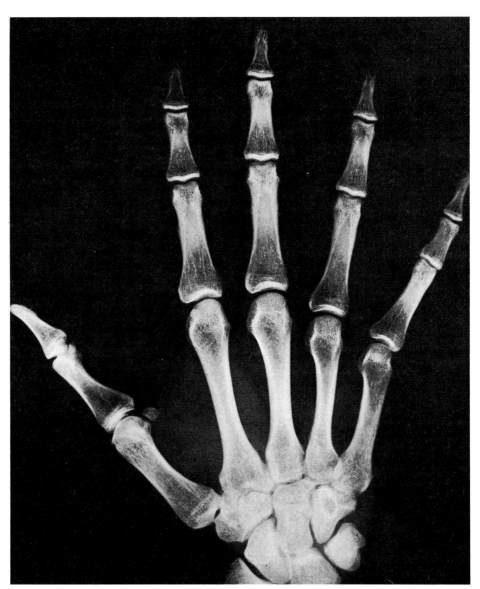

6.1.12

Seminar 2

Attachment of the upper limb to the trunk

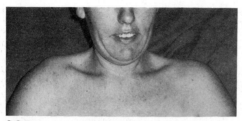

6.2.1

6.2.2

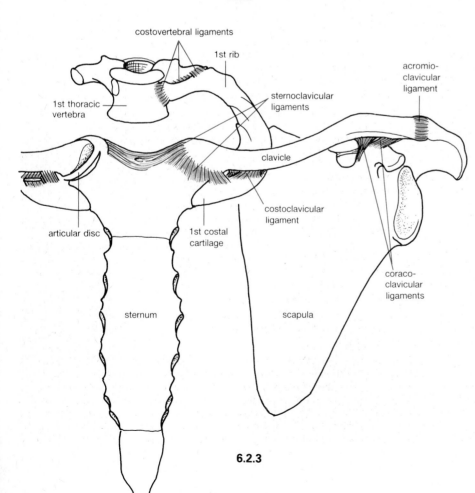

6.2.3

The **aim** of this seminar is to study the way in which the upper limb is attached to the torso; to examine the joint between the sternal end of the clavicle and the manubrium of the sternum (sterno–clavicular); the joint between the acromial end of the clavicle and the acromial process of the scapula (acromio–clavicular); the muscles which attach the clavicle and scapula to the neck; the muscles which attach the clavicle and scapula to the thorax; and the muscles which are attached between the upper end of the humerus and the chest wall.

A. Living anatomy

a) The sterno–clavicular joint is a synovial joint of the plane variety and can therefore glide in all directions. Feel your partner's joint and get him to raise and lower this shoulder and to move it forwards (protraction) and backwards (retraction). Record the movements that you feel at the sterno–clavicular joint. And repeat this on yourself.

b) the acromio–clavicular joint is also a synovial joint of the plane variety. Palpate this joint on your partner having first examined the articulated skeleton.

Qu. 2A *In a dislocation of this joint (6.2.1) which of the two bones will lie uppermost?*

Qu. 2B *The clavicle is a commonly fractured bone. During what sort of accident would it be subjected to a major compression stress along its length?*

Qu. 2C *6.2.2 is a radiograph of a fractured clavicle; What forces would cause the two ends of the bone to take up the positions in which they lie?*

Next examine the movements of the scapula whilst palpating the inferior angle and spine. The scapula is retracted as the shoulders are braced back, and is protracted around the chest wall as the arm pushes forwards. The scapula can be elevated or depressed as the shoulders are raised or lowered. Finally, examine the rotation of the scapula that occurs as the arm is raised above the head.

B. Prosections

a. Attachment of the clavicle and scapula to the torso (6.2.3)

On a prosection of the sterno–clavicular joint examine:

The fibrous capsule which unites the two bones, and its lining of synovial membrane.

The ligament which strengthens the capsule

superiorly is the **superior sterno–clavicular ligament**: it is also known as the interclavicular ligament because it extends across the midline to the joint on the opposite side; and the anterior and posterior ligaments which strengthen the capsule.

The fibro–cartilaginous **intra–articular disc**: which is attached around its margin to the capsule, while its lower part is tucked under the medial end of the clavicle to be attached to the first costal cartilage.

The subclavius muscle: it attaches the under surface of the clavicle to the 1st costal cartilage, and the short, tough **costo–clavicular ligament** which attaches the sternal end of the clavicle to the 1st costal cartilage.

The extremely strong **coraco–clavicular ligaments**: they attach the under surface of the distal part of the clavicle to the coracoid process of the scapula.

b. Muscles of the back

Examine the attachment of the upper fibres of the **trapezius (6.2.4)**, which arise not only from the external occipital protuberance of the skull and a bony ridge which runs laterally from it (superior nuchal line), but also from the spines of the cervical vertebrae via the thick ligamentum nuchae; the fibres are inserted into the inner aspects of the lateral end of the clavicle and the acromial process, and the upper aspect of the spine of the scapula.

Face your partner and placing your palm on both his shoulders, get him to shrug; feel the contraction of the upper fibres of trapezius. The middle fibres of this muscle arise from the spinous processes of the upper thoracic vertebrae and are inserted into the upper aspect of the spine of the scapula. You will feel these fibres contracting if you get your partner to brace back his shoulders. The lower fibres of the muscle arise from the spinous processes of the lower thoracic vertebrae and pass upwards to be inserted into the upper aspect of the spine of the scapula more medially. Because of the direction of the latter fibres, one of the actions of the muscle is to help rotate the scapula on the chest wall so that the glenoid fossa of the scapula faces upwards. Get your partner to raise his arm straight above his head and feel the rotation of his scapula as he does so.

Examine the prosected part and turn the trapezius laterally noting the muscles, rhomboid major and minor, which attach the vertebral border of the scapula to the spines of the upper thoracic vertebrae. Note too the muscle attached to the supero–medial angle of the scapula, levator scapulae, which arises at its upper end from the posterior aspects of the transverse processes of the upper four cervical vertebrae.

Qu. 2D *What is the action of the rhomboid muscles?*

Examine the **latissimus dorsi (6.2.4)** which arises from thick lumbar fascia attached to the spines of the lumbar and sacral vertebrae, the posterior 1/3rd of the iliac crest and the spines of the lower six thoracic vertebrae. The muscle passes upwards

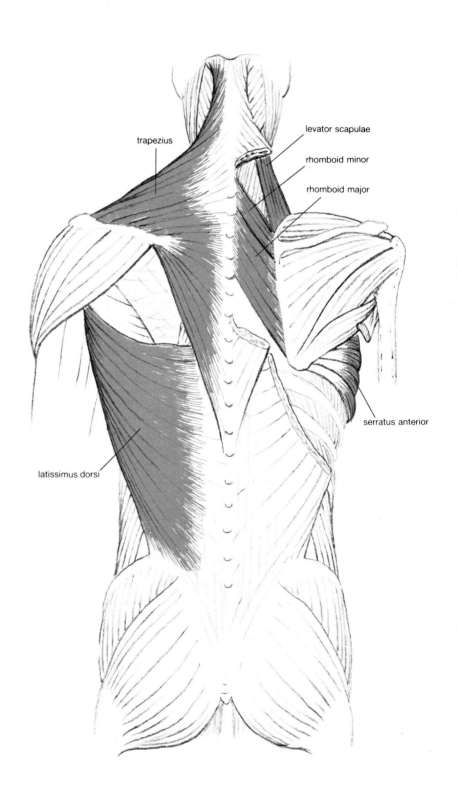

trapezius

levator scapulae

rhomboid minor

rhomboid major

serratus anterior

latissimus dorsi

6.2.4

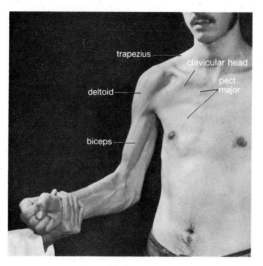

6.2.5

over the angle of the scapula (to which a bundle of muscle fibres is often attached) to be inserted, via a spiral aponeurosis, into the intertubercular groove of the humerus.

Qu. 2E *What is the action of latissimus dorsi?*

c. Muscles of the anterior and lateral aspect of the thorax

Examine the attachments of **pectoralis major** (**6.2.5**, **6.2.6**) to the side of the body of the sternum, the upper six costal cartilages and the anterior aspect of the medial end of the clavicle. Its fibres spiral to be inserted into the lateral border of the intertubercular groove of the humerus.

Qu. 2F *What are the actions of this muscle?*

Examine **pectoralis minor** (**6.2.6**) which lies beneath the pectoralis major, it arises by slips of muscle from the anterior aspect of the 3rd, 4th, and

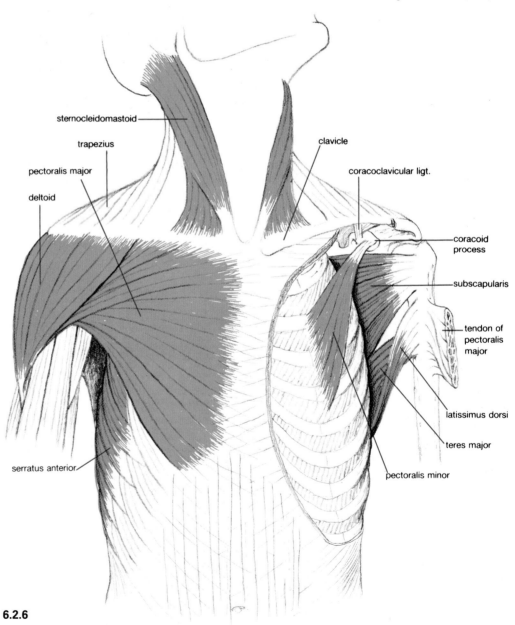

6.2.6

5th ribs to be inserted into the tip of the coracoid process of the scapula.

Qu. 2G *What is the action of this muscle?*

Get your partner to bend his head forward against the resistance of your hand. Note the contraction of a muscle, **sterno–mastoid** (sternocleidomastoid), which arises from the manubrium of the sternum and the sternal end of the clavicle; it is inserted into the bony protuberance of the skull immediately behind the ear (mastoid process of the temporal bone) and the occipital bone behind.

Qu. 2H *How can you get the sterno–mastoid of one side only to contract? What movement takes place if both muscles were to contract?*

Examine the **serratus anterior** (**6.2.6**, **6.2.7**) which arises by individual slips (digitations) from each of the upper eight ribs; these bands of muscle

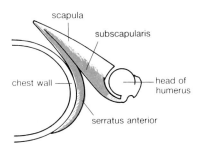

6.2.7 Transverse section of chest wall and scapula.

unite and pass backwards around the lateral wall of the chest to be inserted into the whole length of the vertebral border of the scapula on its thoracic aspect.

The Axilla

Place the flat of your hand on serratus anterior and slide your fingers upwards beneath pectoralis major and minor into the pyramidal shaped space called the **axilla**. Note that pectoralis major and minor lie in front of your hand, forming the **anterior wall** of the axilla (**6.2.6**). Subscapularis and teres major, with the aponeurosis of latissimus dorsi spiralling around it, form the **posterior wall** of the axilla. The **medial wall** of the axilla is formed by serratus anterior with its

nerve lying on the upper part of the thoracic cage.

Neural elements of the brachial plexus arising in the neck pass downward between the clavicle (anteriorly) and scapula (posteriorly); at the first rib they are joined by the first thoracic nerve root, and the subclavian (later termed axillary) artery and vein, all of which have emerged from the thorax. Together the entire **neurovascular bundle**, ensheathed in fascia continuous with that in the neck, crosses the first rib to enter the **apex** of the axilla where it is further surrounded and protected by fat and lymph glands.

Grip, in turn, the lower border of the anterior and posterior walls of your own axilla. The floor of the axilla is dome-shaped because suspensory bundles of fascia originating from that covering pectoralis minor are inserted into the skin and superficial fascia of the axillary floor.

It was pointed out earlier that the clavicle acts as a strut, holding the shoulder joint away from the chest wall. This ensures a considerable degree of mobility of the upper limb. The medial end of the clavicle, as you have seen, articulates with the manubrium of the sternum, and is held down by the strong costoclavicular ligament. Its lateral end is suspended from the skull and cervical spine by the upper fibres of the trapezius and itself suspends the arm by the coraco–clavicular ligament and muscles.

Movements of the lateral end of the clavicle occur upwards, downwards, backwards and forwards, pivoting on its medial ligamentous attachment. When fractures occur they do so mostly in the middle 1/3rd of the bone and are due to falls on the shoulder or outstretched arm when the weight of the body is transmitted along the shaft of the bone.

When you have completed seminars 9–12 on the innervation of the upper limb, make a note of the nerves which supply the muscles you have been studying.

Requirements:
An articulated skeleton
Prosections of the sterno–clavicular joint; cora-
coclavicular ligaments; superficial and deep
muscles of the back; muscles of the anterior
and lateral aspects of the chest.
Radiographs of intact end of
fractured clavicle; acromio-clavicular joint;
sterno-clavicular joint.

Seminar 3

Shoulder joint

The **aim** of this seminar is to study the shoulder joint and its movements.

The shoulder joint comprises a shallow articulation between the head of the humerus and the glenoid fossa of the scapula; it is a synovial joint of the ball and socket variety which, with the movements of the scapula, provides an exceedingly wide range of movements.

A. Living anatomy

Explore the range of possible movements at the shoulder joints on yourself and your partner: abduction and adduction; flexion and extension; medial and lateral rotation—a combination of all these movements comprises circumduction of the arm. Perform all these movements again, at the same time palpating the anterior aspect of the head of the humerus. Make certain that you can feel the greater and lesser tubercles of the humerus and the inter–tubercular groove as you rotate the arm.

Qu. 3A *When movements are occurring at the shoulder joint are movements also occurring at other joints?*

Qu. 3B *Can all the movements at the shoulder joint be performed if the scapula remains fixed?*

The contour of the normal shoulder, when looked at from in front, is formed from above downwards by the trapezius, the acromion, and the acromio–clavicular joint and the deltoid overlying the greater tubercle of the humerus. The deltoid also has a forward convexity due to the underlying greater and lesser tubercles of the humerus. Medial to this anterior prominence (where a depression can be seen), the coracoid process of the scapula can be felt just below the clavicle.

B. Radiographs

Examine radiographs which show the joint in various positions. It is important to note that the articular surfaces of the humeral head and the glenoid fossa lie parallel to one another (see **6.1.4**).

C. Prosections

a. The shoulder joint

Examine the prosected specimen and note the shallow **glenoid fossa** which is covered by **hyaline cartilage**, and deepened around its rim by a **glenoid labrum** of articular fibrocartilage. Note that the articular surface covers only a portion of the almost spherical **head of the humerus**.

Examine the **capsule** (**6.3.1**) which is attached around the cartilaginous rim of the glenoid fossa

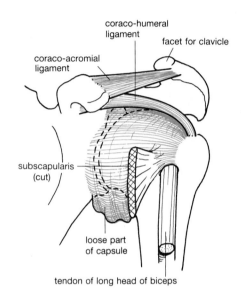

coraco-humeral ligament

coraco-acromial ligament

facet for clavicle

subscapularis (cut)

loose part of capsule

tendon of long head of biceps

6.3.1

and to the anatomical neck of the humerus except inferiorly where the attachment passes downwards onto the medial aspect of the shaft. The capsule is loose inferiorly which, with the shallow articulations, is commensurate with a wide range of movement. The joint is adapted for mobility rather than stability. Note the glistening **synovial membrane** which, in any synovial joint, lines the capsule and all those parts of the joint which are not covered by hyaline articular cartilage. Its function is to produce **synovial fluid**.

Qu. 3C *What are the functions of synovial fluid? What is meant by its thixotropic properties?*

Examine the rather insubstantial thickening or **intrinsic ligaments** of the capsule. These can be seen anteriorly (**gleno–humeral ligaments**) passing from the anterior aspect of the glenoid fossa to the neck of the humerus, and superiorly (**coraco–humeral ligament**) passing from the base of the coracoid process onto the tuberosities of the humerus. Both anteriorly and posteriorly the capsule gains added strength from the attachments of the muscles, **'rotator cuff' muscles**, which are closely associated with the joint. The weakest and least protected aspect of the capsule lies inferiorly.

Extrinsic ligaments are not thickenings of the capsule (**6.3.2**). The **coraco–acromial ligament** is triangular in shape and as its name suggests its base is attached to the coracoid process while its apex passes to the tip of the acromion. It protects the superior aspect of the joint, thereby preventing upward dislocation of the humerus. Beneath this ligament is a large subacromial bursa which does not open into the shoulder joint. The **long tendon of the biceps** is attached to a tubercle at the top of the glenoid fossa (supraglenoid tubercle) and passes through the joint before giving rise to its muscular belly; this tendon helps to stabilize the head of the humerus within the joint when movements are taking place.

Qu. 3D *In which direction does a dislocation of the head of the humerus usually occur?*

Qu. 3E *Which nerve is most likely to be damaged in a dislocation of the shoulder joint? (see Seminar 12)*

In general, any artery lying close to a joint will supply it and its surrounding musculature. Nerves supplying muscles also supply the joints over which those muscles act. This provides the basis for a protective reflex which protects the joint against overstretching of its capsule. A nerve supply to a joint is essential since a) sensory nerve endings in the capsule and synovial membrane and in adjacent muscles and tendons provide information as to the position of the joint in relation to space (proprioception); and information with regard to pain which helps to prevent excessive or damaging movements at the joint; b) vasomotor fibres of the sympathetic nervous system supply arterioles of the synovial membrane and regulate the production of synovial fluid.

b. Shoulder muscles

Examine the following muscles:

Subscapularis (6.3.3) arises from the subscapular fossa and is inserted into the lesser tuberosity of the humerus, it is also inserted into and reinforces the capsule anteriorly. Note that it forms a major part of the posterior wall of the axilla.

Qu. 3F *What are the actions of this muscle?*

Qu. 3G *Which muscles other than subscapularis form the posterior wall of the axilla?*

Teres major arises from the dorsum of the scapula near the lower third of the lateral border, and passes laterally in front of the neck of the humerus to be inserted into the medial lip of the intertubercular groove of the humerus just below subscapularis (see **6.2.6**).

Supraspinatus (6.3.4) arises from the supraspinous fossa of the scapula and runs directly above the shoulder joint to be inserted into the capsule and the top of the greater tuberosity of the humerus.

Qu. 3H *What is the major action of this muscle?*

Qu. 3I *Which other muscles will act in combination with it?*

Infraspinatus (6.3.4) arises from the infraspinous fossa of the scapula and is attached to the capsule and the greater tuberosity of the humerus between the supraspinatus and the teres minor.

Teres minor (6.3.4) arises from the dorsal aspect of the lateral border of the scapula and is inserted into the back of the greater tuberosity of the humerus below infraspinatus.

Qu. 3J *What are the combined actions of infraspinatus and teres minor?*

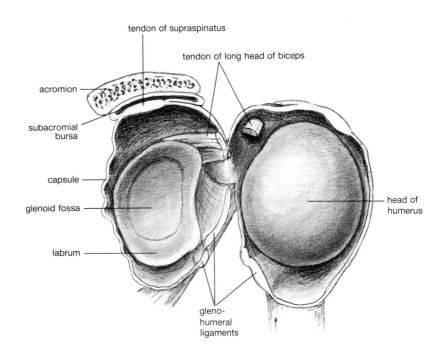

6.3.2 Exposure of interior of shoulder joint.

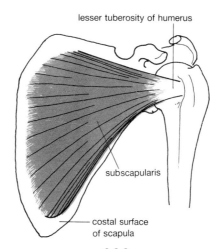

6.3.3

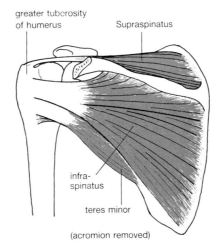

6.3.4

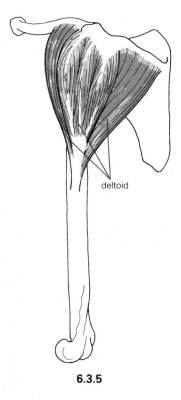

deltoid

6.3.5

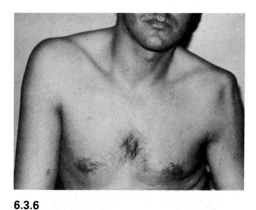

6.3.6

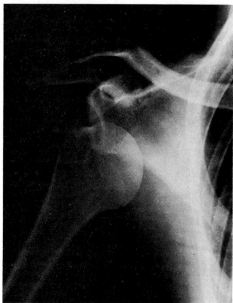

6.3.7

Deltoid (**6.3.5** and see **6.2.5**) resembles an epaulette and arises from the outer aspect of the lateral one third of the clavicle, the acromion process and the spine of the scapula; it is inserted into the deltoid tuberosity half way down the lateral aspect of the shaft of the humerus.

Qu. 3K *What is the action of this muscle?*

Particular attention should be paid in this prosection to the way in which the 'rotator cuff' muscles (subscapularis, supraspinatus, infraspinatus, and teres minor) are blended together and their deeper parts fuse with the capsule of the shoulder joint to form the tendinous 'rotator cuff'. Because of the great mobility of the shoulder joint, its capsule is necessarily lax, and the articular surfaces potentially unstable. But, due to the reinforcement of the capsule by the tendons of the rotator cuff muscles which are actively contracting in movements of the shoulder, the joint is stable in any position.

6.3.6 & **6.3.7** are a photograph and radiograph respectively of a young man who hurt himself playing rugby football. His left shoulder is very painful and he cannot move it. Instead of the normal smooth curve from acromion to deltoid there is a rather sharp angle. The deltoid lacks its normal bulging appearance. It looks 'empty', with one or two shallow dimples. There is a bulge (where the normal shoulder shows a depression) below the point at which the coracoid process should be palpable. Instead of the coracoid process, a large hard lump is to be felt at this site.

Qu. 3L *What has happened to this patient?*

After completing seminars 9–12 make a note of the nerves supplying the muscles you have been studying.

Requirements:
    Articulated skeleton and separate bones of the shoulder
    Prosections of the shoulder joint with capsule intact and capsule opened
    Prosections of the muscles of the shoulder
    Synovial fluid (obtainable from butcher's shop or slaughterhouse).
    Radiographs showing joints in various positions; and centres of ossification.

Seminar 4

Elbow joint

The **aim** of this seminar is to study the elbow joint. It is a synovial joint of the hinge variety which enables the forearm to be flexed and extended on the arm.

A. Living anatomy

Note that on moving your elbow joint movements are limited to flexion and extension. Any muscle which is attached to bones of the shoulder or to the humerus, and which crosses the elbow joint anteriorly to insert into the bones of the forearm, must necessarily flex the joint. Conversely, those muscles which cross the posterior aspect of the elbow must extend the joint. Examine muscles of the anterior and posterior compartments by flexing and extending the elbow against resistance (see **6.2.5**). Note the position of the lateral and medial epicondyles and the olecranon process when the elbow is bent; they form the corners of a roughly equilateral triangle.

B. Radiographs

Examine the radiographs of the elbow joint. Note the position of the bones in the extended (**6.4.1**—anteroposterior view) and flexed (**6.4.2**—lateral view) position of the bones and revise the secondary centres of ossification (Seminar 1). In **6.4.2** draw a line along middle of the shaft of the radius through the centre of the head of the bone. Note that it passes through the centre of a circle which is outlined in part by the capitulum. In certain injuries, for instance a fracture of the shaft of the ulna, the head of the radius may be displaced from its position against the capitulum (**6.4.3**).

Draw a second line down the anterior border of the shaft of the humerus (**6.4.2**). This too should pass through the centre of the circle referred to above. The lower end of the humerus is normally

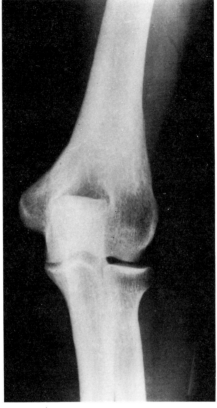
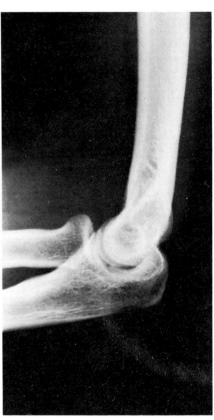

6.4.1   **6.4.2**

angled 45° forwards with respect to the shaft of the bone. If the lower end is displaced it is probably due to a supracondylar fracture of the shaft.

Note the angle between the axis of the humerus and that of the forearm. This 'carrying angle' is greater in women than in men (see p. 20).

Qu. 4A *Examine **6.4.4**: what abnormality can you see?*

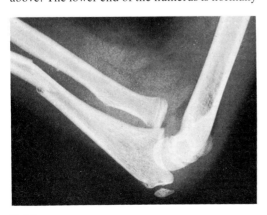

6.4.3

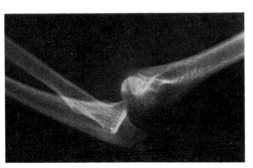

6.4.4

C. Prosections

a. The elbow joint
(6.4.5–6.4.10)

Examine the prosected elbow joint and note the shapes of the articular surfaces which are covered by hyaline articular cartilage; they consist of the quadrilateral–shaped **trochlea of the humerus** which articulates with the deep **trochlear fossa of the ulna**, and the **rounded capitulum** at the lower end of the humerus, which articulates with the rounded concave upper aspect of the **head of the radius**. The articulation between the upper end of the radius and ulna is termed the **superior radio–ulnar joint** and will be considered again later. Note that the head of the radius also articulates on its medial aspect with the radial notch on the upper lateral aspect of the ulna. Holding the head of the radius against the ulna is a sling–type **annular ligament** which is attached to either side of the radial notch. This ligament

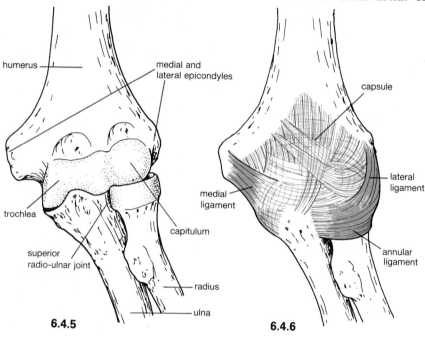

6.4.5

6.4.6

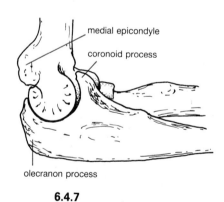

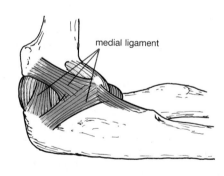

6.4.7

6.4.8

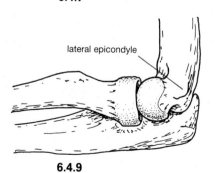

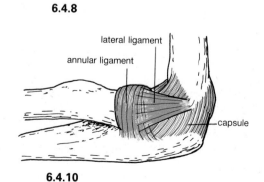

6.4.9

6.4.10

enables the head of the radius to rotate when its lower end is rotating around the lower end of the ulna (pronation and supination of the forearm).

Examine the **capsule** which is common to both the elbow and superior radio-ulnar joints. It is attached to the humerus immediately above the coronoid and radial fossae and to the inferior aspects of the medial and lateral epicondyles. Posteriorly it is attached to the upper margin of the olecranon fossa so that, when the elbow is straightened, the olecranon process of the ulna fits snugly into the fossa. Inferiorly, the capsule is attached to the annular ligament of the head of the radius so that movements of the upper end of the radius are not impeded; on the ulna, the capsule is attached to the rim of the trochlear notch.

The **synovial membrane** lines the capsule and all those parts of the joint which are not covered with hyaline articular cartilage; it also projects beneath the annular ligament of the head of the radius to surround the neck of the bone. As you might expect, the capsule of a hinge joint is reinforced medially and laterally by **intrinsic, collateral ligaments**. The **lateral ligament** is fan-shaped and is attached to the lateral epicondyle superiorly and to the annular ligament inferiorly. The **medial ligament** is triangular in shape with three thickenings which are attached, in order, to the medial epicondyle superiorly, the coronoid process of the ulna inferiorly, and the medial side of the olecranon process.

b. Muscles of the anterior compartment of the arm

In the upper limb sheets of fascia (intermuscular septa) extend from the deep fascia, between the major groups of muscles, to the lateral and medial sides of the shaft of the humerus. In the forearm similar sheets of fascia pass from the deep fascia to the lateral and medial sides of the shafts of the radius and ulna respectively. This arrangement serves to separate groups of muscles and enables them to obtain a wider origin, not only from bone but also from fascia. It is convenient, for purposes of description, that these sheets of deep fascia divide the limb into anterior and posterior compartments. Furthermore this reflects the basic embryological division of limb musculature into flexor and extensor groups.

Examine the muscles which lie in the anterior compartment of the arm:
Biceps (6.4.11) arises from the scapula by two heads, a **short head** from the tip of the coracoid process, and a **long head** from the supraglenoid tubercle. The tendon of the long head lies within the capsule of the shoulder joint for some distance. The combined heads form a large muscle belly which is inserted, by a flattened tendon passing in front of the elbow joint, into the posterior part of the tuberosity of the radius.

An expansion of the tendon, the **bicipital aponeurosis**, crosses medially over the forearm superficial flexor muscles to be attached to the posterior border of the ulna via the deep fascia. In this way biceps exerts its flexor action on both bones of the forearm. In the 'position of function' biceps is a strong supinator and flexor of the forearm (see **6.2.5**); it contracts strongly when a screw is being driven home or when a cork is being removed from a bottle. Since the tendons of origin cross the shoulder joint the muscle also has a flexor action at this joint, while the long tendon helps to stabilise the head of the humerus when movements of the shoulder are taking place.

Brachialis (6.4.12) arises from the lower half of the front of the humerus and is inserted into the front of the coronoid process of the ulna.

Coraco–brachialis (6.4.12) is found in the upper arm and arises, along with the short head of biceps, from the tip of the coracoid process; it is inserted into the medial aspect of the humerus half way down the shaft.

Qu. 4B *What are the actions of brachialis and coraco–brachialis?*

c. Muscles of the posterior compartment of the arm

The massive **triceps** muscle (**6.4.13**), as its name implies, arises from three separate heads, a **long head** from the infraglenoid tubercle of the scapula which lies outside the capsule of the shoulder joint, a **lateral head** from the posterior aspect of the shaft of the humerus above the radial groove and a **medial head** which takes origin from the broad surface below the groove. The **anconeus**, really a part of triceps, continues the muscle from the lateral epicondyle down to the lateral side of the olecranon process. The three heads of the muscle unite to be inserted into the upper surface of the olecranon process of the ulna.

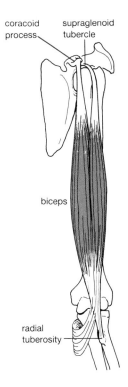

coracoid process

supraglenoid tubercle

biceps

radial tuberosity

6.4.11

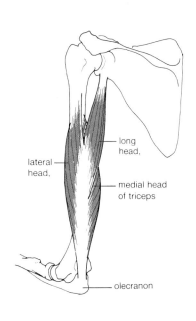

lateral head,

long head,

medial head of triceps

olecranon

6.4.13

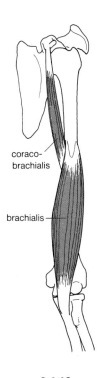

coraco-brachialis

brachialis

6.4.12

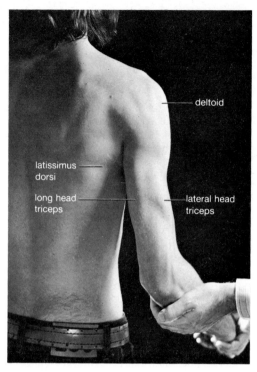

6.4.14

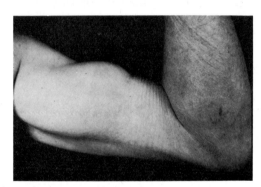

6.4.15

Qu. 4C   *What are the actions of triceps?*

Flex your pronated forearm against resistance and then repeat the movement with the forearm in a supinated position.

Qu. 4D   *In which of these movements does biceps play a part?*

With the elbow bent, extend the arm against resistance as in **6.4.14**.

Qu. 4E   *Which group of muscles is contracting most strongly during this movement?*

With the elbow semiflexed and supported, and the hand holding a weight, gradually extend the elbow joint.

Qu. 4F   *Which group of muscles is contracting most strongly?*

Before leaving the prosections notice again the long head of biceps. Its tendon runs a prolonged course from the supraglenoid tubercle of the scapula over the head of the humerus within the shoulder joint to emerge beneath the capsule in the intertubercular groove. It then expands into its muscle belly, which often remains separate from the short head, until near the elbow. Within the shoulder joint it is enclosed within a synovial sheath, which sometimes becomes inflamed, causing pain. The tendon is also prone to spontaneous rupture due to degeneration of collagen within the tendon. Following a strain, which is not necessarily severe, the patient experiences a sudden pain in the front of the upper arm and a few days later some bruising appears. The contour of the biceps is altered. An example of this injury is seen in **6.4.15** but there is surprisingly little dysfunction since the short head is quite strong. There may be some weakness of supination of the forearm, but flexion of the elbow can still be performed by brachialis.

Qu. 4G   *With your knowledge of the position of the bony prominences around the elbow, how could you distinguish between a supracondylar fracture of the humerus and a dislocation of the ulna—without recourse to radiography?*

When you have completed seminars 9–12 make a note of the nerve supply to the muscles you have been studying.

Requirements:
  Articulated skeleton and separate bones of the
    upper limb
  Radiographs of the elbow joint in adults and
    children
  Prosections of the elbow joint, muscles of the
    anterior and posterior compartments of the
    upper arm.

Seminar 5

Joints of forearm and hand

The **aim** of Seminars 5 and 6 is the study of the joints and movements of the forearm and hand.

A. Living anatomy

Movements of the forearm which enable the hand to be related with respect to the humerus (pronation and supination) occur at the **superior and inferior radio–ulnar joints**.

Keep your elbows to your sides and bend your arms to a right angle. Now rotate your forearms so that the palms face the floor—the forearms are now in the position of pronation. Next rotate the forearms so that the palms face the ceiling—they are now supinated.

Qu. 5A *Against resistance, is pronation or supination the stronger? Why do screws have right–hand threads?*

Extend your arm and touch a wall in front of you with one extended finger, now pronate and supinate your forearm and do the same with each finger in turn. You will see that the forearm can rotate about an axis that can pass through any finger.

Qu. 5B *How is this possible?*

Now pronate and supinate the forearms freely without contact with the wall.

Qu. 5C *Through which finger does the axis of movement pass in this free movement?*

Look at the photograph of a worker holding a tool (**6.5.1**) and the student holding the pen (**6.5.2**).

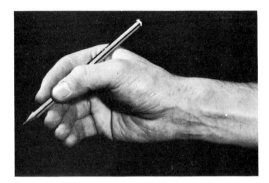

6.5.2

There are obviously numerous 'positions of function' of the forearm/wrist/hand, but some should be particularly noted. We hold heavy tools or exert force on objects using a '**power grip**' and perform more delicate manoeuvres with a '**precision grip**'. Comment on the position of the various joints that you can see in the photographs.

When the forearm has to be immobilized, as after a fracture, this is commonly done with the arm in the semiflexed elbow, mid–prone forearm position. Supinate your forearm against resistance and note the contraction of the biceps. Place your left thumb over the tendon of biceps on the front of your partner's elbow and tap your thumb with a patellar hammer. This should stretch the biceps and give rise to a 'biceps jerk' the quality of which depends on the degree of excitation at the spinal cord level of cervical 5 & 6 and on the integrity of the nerve which supplies the muscle, its neuromuscular junction, and the muscle itself. With your arm, or that of your partner, pronated, place the tips of your fingers over the head of the radius and its immediately subjacent shaft and then supinate the forearm against resistance; you should, with a little practice, be able to feel the contraction of supinator (see **6.6.13**).

With your right arm lying in the prone position and with the elbow flexed hold the lower ends of the radius and ulna with your left hand so that they do not move.

Qu. 5D *What movements can you perform at i) the wrist and ii) the metacarpophalangeal joints?*

Both the superior and inferior radio–ulnar joints are of the synovial pivot variety. Examine the articular surfaces of these joints on the skeleton.

Qu. 5E *Of the range of possible movements of a joint, which movements cannot take place at the joints under discussion?*

6.5.1

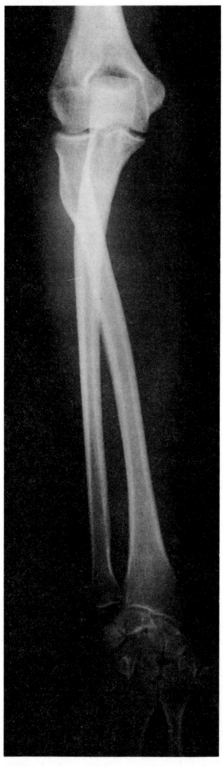

6.5.3

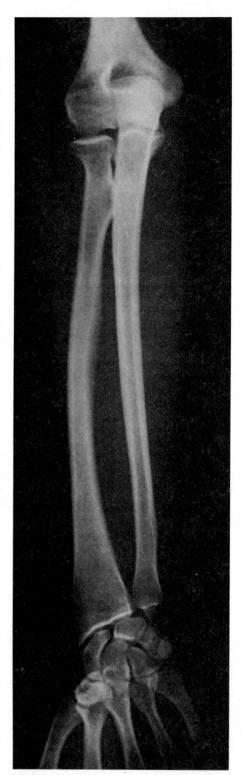

6.5.4

Move your fingers at the phalangeal joints. Examine the articular surfaces on the joints of the skeleton.

Qu. 5F *How would you classify these joints and why?*

Rest the back of your hand against a flat surface with your thumb at right angles to the rest of the

fingers, i.e. with the nail directed laterally. This is the natural position of the hand. Note the degree of flexion of each of the fingers. Now look at the skeleton and examine the concavo–convex shape of the bases of the **1st metacarpal and the trapezium**. This synovial joint is aptly termed a **'saddle joint'** and is unique to Man. Flex the thumb across the palm and then extend it; abduct it at right angles away from the index finger and then adduct it to its original position. Approximate the thumb and little finger pads. This movement of opposition is, with respect to its range, also unique to Man; it is achieved by a combination of movements including that of medial rotation.

Qu. 5G *Why is the ability to oppose the pad of the thumb and little finger of such importance to Man?*

Qu. 5H *During the course of evolution and the development of opposition and other manipulative movements of the hand and fingers, further evolvements relating to the neural control of these movements were taking place. What might these be?*

B. Radiographs

Revise the radiographs of the forearm and hand which you studied in seminar 1. **6.5.3** & **6.5.4** show the forearm in pronation and supination, respectively.

C. Prosections

Superior and inferior radio–ulnar joints, the wrist joint and the joints of the hand and fingers

a) Examine the **superior radio–ulnar joint** (**6.5.5**) between the head of the radius and the radial notch on the upper end of the ulna. It is enclosed within the capsule of the elbow joint. The head of the radius is contained within a cup–shaped **annular ligament** attached to both edges of the radial notch on the ulna. During movements of pronation and supination of the forearm the head can then swivel on the capitulum. If a child's forearm is suddenly pulled e.g. if it falls while holdings its mother's hand, the head of the radius may be drawn distally through the annular ligament.

b) Examine also the **inferior radio-ulnar joint** (**6.5.5**), between the lower ends of the radius and ulna. Both the superior and inferior joints are involved in pronation and supination of the forearm and are together classified as **synovial joints of the pivot variety**.

Fractures of the lower ends of the radius and ulna are very liable to produce distortion of the inferior radio–ulnar joint. Such distortion is likely to cause limitation of movement of the inferior radio–ulnar joint, so that full pronation and supination movements may not be possible. Loss of pronation can interfere with writing, and loss of supination with turning the hand over to receive something, or to open a door.

Note the thick sheet of collagenous tissue which unites the radius and ulna along their adjacent shafts, **the interosseous membrane**.

Examine carefully the **triangular shaped intra–articular disc** (**6.5.6**) which is attached at its apex to the styloid process of the ulna, and at its

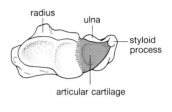

6.5.6

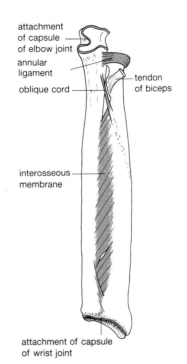

6.5.5

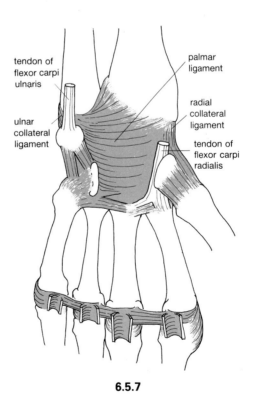

6.5.7

c) The **wrist joint** (**6.5.7**): examine its capsule —proximally it is attached around the articular margins of the base of the radius and ulna while distally it is attached to the proximal row of carpal bones, i.e. the scaphoid, lunate, and triquetral. The attachments of the triangular articular disc prevent the lower end of the ulna from articulating with the carpus. Make a note of the bony articulations at the wrist joint when the hand is abducted and again when it is adducted.

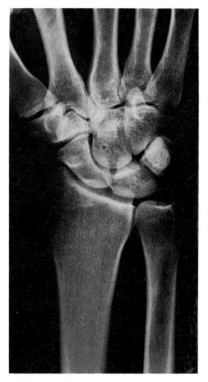

6.5.8

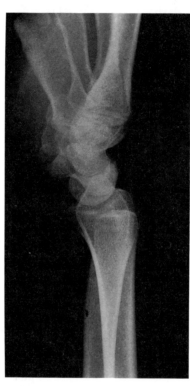

6.5.9

Note the angle of slope of the distal articular surface of the radius in relation to its long axis in both antero–posterior and lateral views (**6.5.8**, **6.5.9**). In the antero–posterior view, if a line is drawn from the tip of the styloid process of the radius to the tip of the styloid process of the ulna, it lies at an angle of about 75° to the long axis of the shaft of the bones, the articular surface facing ulnarwards. Similarly in a lateral view a line drawn joining the margins of the carpal articular surface of the radius commonly slopes about 15° to the long axis of the shaft of the bone, the articular surface facing anteriorly. The articular surface of the radius can be displaced in falls on the wrist and it is important that this is corrected if full movement of the wrist is to be restored.

A Colles fracture, named after the surgeon who described the injury in the years before x–rays were discovered, is the most common fracture occurring at the wrist (**6.5.10**, **6.5.11**). If the dorsal capsule of the wrist joint is opened so that you can view the anterior part of the capsule from within the joint, the strength of the wrist joint is easily appreciated. During a fall on the outstretched hand, especially in the aged (when the bones are relatively less resistant), this strength causes the upward, backward, and lateral displacement of the distal part of the radius which is typical of a Colles fracture.

Make a note of the attachments of the capsular ligaments of the wrist joint, the metacarpo–phalangeal joint and an inter–phalangeal joint.

d) **The metacarpo–phalangeal joints**: in full extension of the metacarpo–phalangeal joints of the index, middle, ring, and little fingers there is a range of side–to–side movements of the fingers described as abduction and adduction. These movements are described in relation to an axis taken through the middle finger, thus abduction is away from, and adduction towards, this axis. When these metacarpo–phalangeal joints are flexed the side–to–side movements are lost. This is because, in lateral profile, the surface of the head of a metacarpal is shaped like a cam (**6.5.12**). The collateral ligaments of the joint pass from the dorsal tubercles of the head of the metacarpal to the lateral tubercles of the proximal phalanx. If the metacarpo–phalangeal joints are extended then these ligaments are lax and therefore abduction and adduction movements are possible. In flexion, the collateral ligaments are tightened thus abolishing any side–to–side motion of the joint. This is obviously beneficial in increasing the stability of grip around a small object in the hand, such as the handle of a tool. However, if a hand is injured, it swells, as does any other part of the body. Fluid accumulates in the tissues and, containing protein, may set, eventually forming fibrous tissue. If an injured hand is immobilized so that the metacarpo–phalangeal joints are extended, the collateral ligaments may shorten in their relaxed state if fibrous tissue is formed. On attempting to flex the metacarpo–phalangeal joints, the patient may then find it impossible to make a satisfactory grip because the collateral ligaments become too taut,

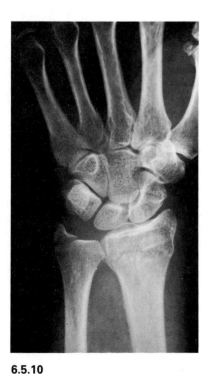

6.5.10

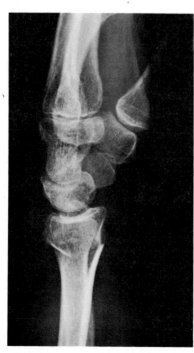

6.5.11

short of full flexion. Therefore, if a hand has to be put in plaster due to injury, the metacarpo–phalangeal joints of the ulnar four digits are inmobilized in nearly full flexion, so that shortening of the collateral ligaments is not possible, and a normal range of movement is obtained when mobilization begins.

e) The **interphalangeal joints**: there is little cam effect in these joints. The collateral ligaments are taut throughout the range of flexion and extension. In general, after injuries, the fingers are best immobilized so that the metacarpo–phalangeal joint is flexed to 80°, and the interphalangeal joints to 10° each.

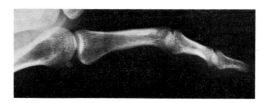

6.5.12

Requirements:
    Articulated skeleton of upper limb
    Prosections of superior and inferior radio–ulnar joints, wrist, metacarpo–phalangeal and interphalangeal joints
    Radiographs of these joints

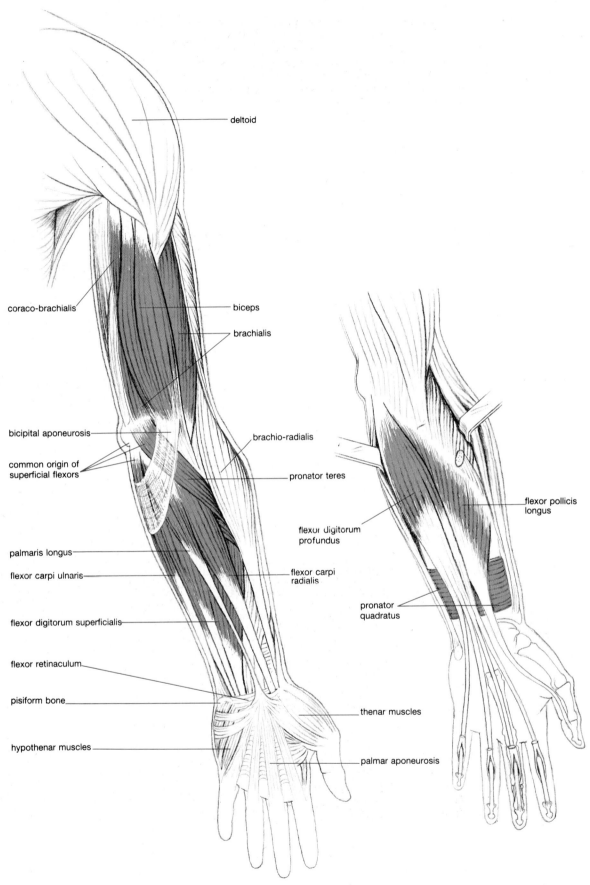

deltoid

coraco-brachialis

biceps

brachialis

bicipital aponeurosis

brachio-radialis

common origin of
superficial flexors

pronator teres

flexor digitorum
profundus

flexor pollicis
longus

palmaris longus

flexor carpi ulnaris

flexor carpi
radialis

flexor digitorum superficialis

pronator
quadratus

flexor retinaculum

pisiform bone

thenar muscles

hypothenar muscles

palmar aponeurosis

6.6.1   **6.6.2**

Seminar 6

Muscles and movements of forearm and hand

When describing the forearm it is often helpful to use the terms 'ulnar' and 'radial' rather than medial and lateral, because pronation can obscure the basic anatomical position.

a. Muscles of the anterior compartment of the forearm

The anterior compartment of the forearm contains flexor muscles of the wrist and fingers, and muscles which produce pronation and supination of the forearm. Some muscles also have an action on the elbow joint by virtue of their attachment to the humerus (**6.6.1**, **6.6.2**).

i) Examine first the **common origin of the superficial flexor muscles of the forearm** which is attached to the medial epicondyle. Identify **pronator teres** (**6.6.3**) which arises by two heads from *a*) the common flexor origin and the lower end of the medial supracondylar ridge and *b*) the medial border of the coronoid process of the ulna; when in a later seminar you come to study the median nerve you will note that, as it leaves the front of the elbow, it passes between the two heads of pronator teres to enter the forearm. The muscle passes diagonally across the forearm to insert halfway down the front of the shaft of the radius; as its name suggests, it pulls the radius across the ulna thus pronating the forearm.

Flexor carpi radialis also arises from the common flexor origin and inserts into the bases of the 2nd and 3rd metacarpal bones.

Palmaris longus, which is not always present, also arises from the common flexor origin and is inserted into the very thick, tough fascia—the palmar aponeurosis, which lies beneath the skin of the palm.

Flexor carpi ulnaris arises not only from the common flexor origin but also from the medial side of the olecranon and the posterior border of the ulna; it is inserted into the pisiform bone with an extension on to the base of the 5th metacarpal.

Flexor digitorum superficialis (**6.6.4**) lies on a deeper plane than the other superficial flexors. It arises from the common flexor origin, from the coronoid process of the ulna, and from the front of the upper part of the shaft of the radius by a thin aponeurosis. Near the wrist the muscle divides to give four tendons. In the fingers these tendons are divided to allow the tendons of flexor digitorum profundus to pass through to their attachments on the bases of the distal phalanges, while the slips from those of flexor digitorum superficialis partially reunite and are inserted into the base and sides of the middle phalanges.

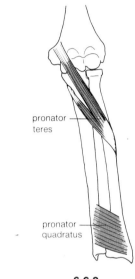

pronator teres

pronator quadratus

6.6.3

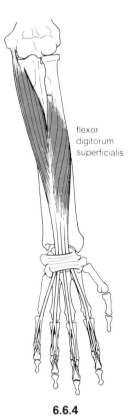

flexor digitorum superficialis

Flex your wrist against resistance and identify the tendons on the anterior aspect of your forearm (**6.6.5**).

Qu. 6A *What are the actions of these individual muscles?*

6.6.4

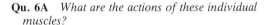

flexor carpi radialis

palmaris longus

flexor carpi ulnaris

6.6.5

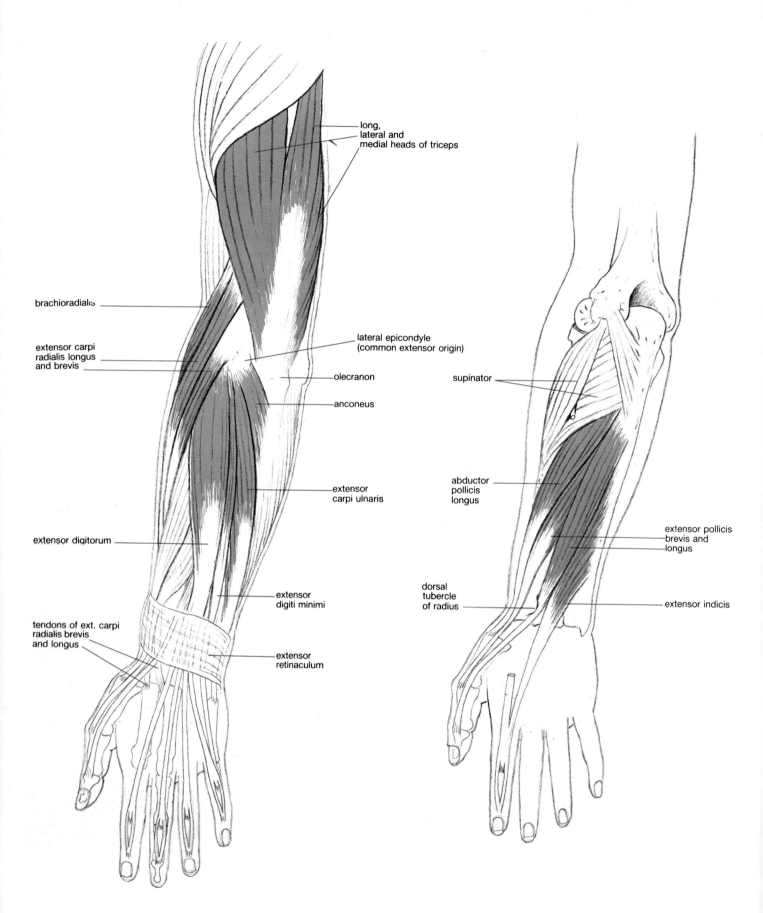

long,
lateral and
medial heads of triceps

brachioradialis

extensor carpi
radialis longus
and brevis

lateral epicondyle
(common extensor origin)

olecranon

anconeus

supinator

extensor carpi
ulnaris

abductor
pollicis
longus

extensor digitorum

extensor pollicis
brevis and
longus

extensor
digiti minimi

dorsal
tubercle
of radius

extensor indicis

tendons of ext. carpi
radialis brevis
and longus

extensor
retinaculum

6.6.7   6.6.8

ii) Now examine the deep muscles of the fore-arm:
Flexor digitorum profundus (6.6.6) lies deeply in the forearm and takes origin from the medial and anterior aspect of the upper shaft of the ulna and the adjoining interosseous membrane. Its tendons cross the palm to pass into the fingers. In the fingers these tendons pass through the divided tendons of flexor digitorum superficialis to pass to their insertion on the base of the distal phalanges. The tendons of the superficial and deep flexors lie within the fibrous flexor sheaths of the digits. These are strong over the shafts of the phalanges, but more flexible over the joints. They allow the long flexor tendons to exert their pull without 'bow–stringing' away from the phalanges.
Flexor pollicis longus (6.6.6) arises from the anterior surface of the radius between the attachments of pronator teres and pronator quadratus. This strong muscle forms a tendon which receives fleshy fibres down to the wrist, passes deeply through the carpal tunnel and inserts into the base of the distal phalanx of the thumb.
Pronator quadratus (6.6.3) is attached to the lower ends of the shafts of the radius and ulna and, with pronator teres, produces pronation of the forearm.

> **Qu. 6B** *When the forearm is being flexed would you expect muscles involved with pronation and supination to be acting at the same time (i.e. synergistically)?*

Lacerations of the palm of the hand can result in division of both superficial and deep tendons to the fingers. Sometimes the diagnosis of such an injury can be made on sight of the hand. If the normal hand is laid on the table palm upwards, in a relaxed manner, one observes what is described as the 'cascade' of the digits. The index finger is extended most, followed in turn by the remaining fingers, to the little finger, which is extended least.

b. Muscles of the posterior compartment of the forearm (6.6.7, 6.6.8)

i) the posterior compartment of the forearm contains the extensor musculature which acts on the wrist and fingers, and supinator. Many superficial extensors arise from a **common extensor origin** on the anterior aspect of the lateral epicondyle of the humerus.
Brachioradialis (6.6.9) arises from the upper part of the lateral supracondylar ridge and passes down the radial border of the forearm, overlapping both flexor and extensor compartments, to an insertion on the lateral aspect of the distal end of the radius. It is a moderately strong flexor of the elbow, and would tend to bring the forearm from either extreme supination or pronation to the mid–prone position ('position of function').
Extensor carpi radialis longus arises from the lateral supracondylar ridge of the humerus. It is inserted into the dorsal aspect of the base of the 2nd metacarpal.
Extensor carpi radialis brevis arises from the common extensor origin and is inserted into the dorsal aspect of the base of the third metacarpal.
Extensor carpi ulnaris arises from the common extensor origin and is inserted into the dorsal aspect of the base of the 5th metacarpal. These latter three muscles are powerful extensors of the wrist and act synergistically with the long flexors of the fingers. Note the symmetry of distal attachment of the carpal flexors and extensors: when they contract together, the wrist is fixed so that delicate movements of the fingers can occur from a stable base.

> **Qu. 6C** *Get your partner to flex his elbow and then grasp his wrist firmly in order to resist movement. Note the position of your own wrist and hand and comment on your observations.*
>
> **Qu. 6D** *Which muscles of the arm are acting when a glass is raised to the lips?*

Now examine the other superficial extensor muscles of the forearm which arise from the common extensor origin.

Find the **extensor digitorum** and the **extensor digiti minimi**. The tendons of these muscles pass down the back of the forearm and over the back of the hand into the fingers. The tendons on the back of the hand are usually interconnected. Place your hand palm downwards on a flat surface and extend each of your fingers in turn, you will then be able to locate your own tendinous interconnections. Note from the prosection that the tendons are inserted into **fibrous dorsal extensor expansions** lying on the dorsal aspects of the proximal phalanges. These extensor expansions are each inserted by a central slip into the dorsal surface of the base of the middle phalanx, and by two conjoined lateral slips into the dorsal surface of the base of the terminal phalanx. Some small muscles of the palm of the hand (lumbricals and interossei) gain attachment to the extensor tendon expansions. If an extensor tendon is torn from its insertion to the distal phalanx, the condition of 'mallet finger' results **(6.6.10)**. The terminal interphalangeal joint is flexed about 45° and although it will flex, both actively and passively, and extend passively, the patient cannot straighten the joint actively. The injury follows a passive flexion force on the distal interphalangeal joint when the extensor tendon is under strain.

Degenerative changes can arise in or near the common extensor origin. Sharp local tenderness develops in the muscle mass just in front of the lateral epicondyle, and pain is caused when the wrist and fingers are actively extended against resistance. The lesion has been called 'tennis elbow' from its supposed precipitation by back-hand strokes, but in fact very few sufferers seem to play this sport.

ii) Examine the deep extensor muscles of the forearm which are attached largely to the ulna and the interosseous membrane and pass diagonally across the back of the forearm to be inserted into the dorsum of the bones of the thumb and index finger **(6.6.8)**.

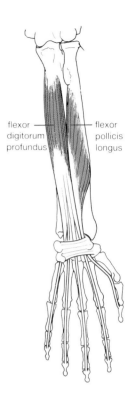

flexor digitorum profundus — flexor pollicis longus

6.6.6

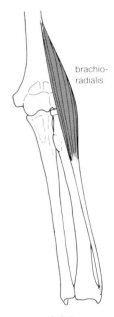

brachio-radialis

6.6.9

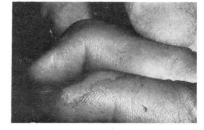

6.6.10

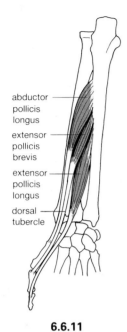

abductor
pollicis
longus

extensor
pollicis
brevis

extensor
pollicis
longus

dorsal
tubercle

6.6.11

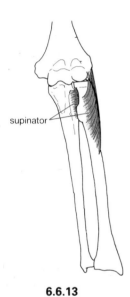

supinator

6.6.13

extensor pollicis
longus

extensor pollicis brevis
and abductor pollicis longus

6.6.12

Abductor pollicis longus (**6.6.11**) is inserted through its long tendon into the base of the 1st metacarpal while **extensor pollicis brevis** is inserted into the base of the proximal phalanx of the thumb. The tendon of **extensor pollicis longus** hooks around the dorsal (Lister's) tubercle of the radius, to be inserted into the base of the terminal phalanx of the thumb. If you extend your thumb, the tendons of these muscles will be clearly seen (**6.6.12**). In fact they form a hollow at the base of the thumb into which snuff can be placed and inhaled—hence the time–honoured name of the 'anatomical snuffbox'. In its depths you should be able to feel the pulsations of the radial artery which winds around the wrist before passing between the heads of the 1st dorsal interosseous muscle to enter the palm.

Spontaneous rupture of extensor pollicis longus tendon may occur and is thought to be due to ischaemia resulting from diseased interosseous vessels. The rupture occurs in the distal part of the tunnel beneath the extensor retinaculum as the tendon changes direction around the dorsal tubercle of the radius. The patient feels that the thumb has 'dropped'; the interphalangeal joint of the thumb will not extend properly and it may get in the way of the fingers.

Deep in the upper part of the forearm locate **supinator** (**6.6.13**). This muscle surrounds the upper third of the radius. It arises from the lateral epicondyle of the humerus, the lateral border of the upper part of the ulna (the supinator crest) and from adjacent ligaments. The ulnar fibres pass behind the radius, join the deep aspect of the humeral fibres and insert into the proximal third of the anterior aspect of the radius, just proximal to pronator teres. The deep branch of the radial nerve gains access to the posterior compartment of the forearm between the two layers.

When you have completed seminars 9–12 make a note of the nerve supply to the muscles you have been studying.

Control of the long tendons at the wrist

The **flexor and extensor retinaculae** are specialized bands of thickened fibrous deep fascia which hold the tendons firmly to the wrist and palm and prevent 'bow-stringing'.

The **flexor retinaculum** is the thick band which converts the concavity of the palmar surface of the carpus into an osteo–fascial channel, the **carpal tunnel**. It is attached to the pisiform and hook of hamate medially, and to the tubercle of the scaphoid and ridge on the trapezium laterally. Mark its position on your hand after palpating these body landmarks.

The superficial and deep flexor tendons to the thumb and fingers pass through the carpal tunnel covered with synovial sheaths together with the tendon of flexor carpi radialis and the median nerve (**6.6.14**).

The **extensor retinaculum** stretches across the back of the wrist and converts the grooves on the dorsum of the distal end of the radius into separate channels for the long extensor tendons and their synovial sheaths. It is attached medially to the pisiform bone and hook of the hamate (as with the flexor retinaculum) and laterally to the radius.

Qu. 6E *Why is the extensor retinaculum not attached to the dorsal aspect of both the radius and the ulna?*

Requirements:
Articulated skeleton and separate bones of the upper limb
Radiographs of elbow, wrist, and hand
Prosections of the superior and inferior radio-ulnar joints
Prosections of anterior and posterior compartments of the forearm showing *a*) superficial and *b*) deep muscles
Prosections of the flexor and extensor retinaculae

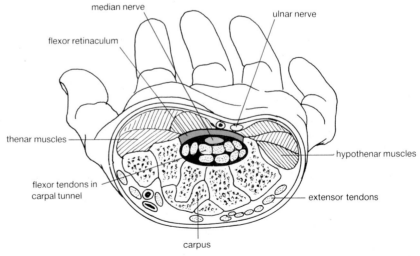

median nerve

ulnar nerve

flexor retinaculum

thenar muscles

hypothenar muscles

flexor tendons in
carpal tunnel

extensor tendons

carpus

6.6.14

Seminar 7

Blood supply and lymphatics of upper limb

The **aim** of this seminar is to examine the arterial supply to the upper limb and its venous drainage; and also to consider the channels along which extracellular fluid, which has not recirculated through the veins, reaches the venous circulation via the lymphatics.

A. Living anatomy

The pulsations of the arteries to the upper limb can be felt (palpated) at certain points along their course where they can be compressed against bone (**6.7.1**).

Feel for the pulsation of the subclavian artery as it crosses the 1st rib by pressing your index finger downwards between the clavicle and the upper border of the scapula. It is possible to compress the artery against the 1st rib if there is severe haemorrhage of the arm but this is uncomfortable, largely because nerve trunks are lying in close proximity to the artery as it enters the axilla.

Place the flat of your hand against the upper and medial aspect of the arm with your fingers as high in the apex of the axilla as possible. Feel for the pulsation of the axillary artery and compress it against the medial aspect of the shaft of the humerus.

Feel for the pulsation of the brachial artery against the mid shaft of the humerus, and also where it lies on the medial side of the biceps tendon as it crosses the front of the elbow. On rare occasions it may be necessary to apply pressure to the brachial artery in the upper arm if severe haemorrhage is occurring from the vessel below, and direct pressure has failed to stop it. The best place to do so is where the artery lies in front of the coracobrachialis, squeezing the artery against the shaft of the humerus.

The brachial artery, at the front of the elbow, is the best site from which to take a reading of the blood pressure: the artery is constricted in the upper arm by a pneumatic tourniquet cuff connected to a manometer; the cuff is inflated until the vessel is occluded and the radial pulse has disappeared. A stethoscope is then applied over the brachial artery on the medial side of the tendon of biceps and the pressure of the tourniquet is released until the first sound of blood flowing through the artery is heard. The pressure indicated by the manometer at this point indicates the systolic blood pressure. As the tourniquet pressure is released further the sound heard in the cubital fossa at first increases and then quite suddenly disappears, when a continuous flow of blood is achieved. The pressure reading at this point represents the diastolic blood pressure.

Feel the pulsation of the radial artery at the wrist. Compress the artery against the flattened lower end of the radius and note the calibre of its walls. Count the pulsations of the artery for the period of a minute. Feel also the pulsations of the radial artery as it passes over the lateral aspect of the wrist under cover of the long tendons passing into the thumb.

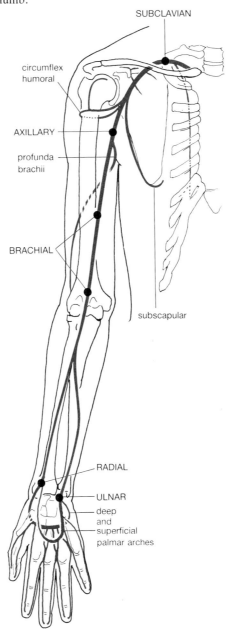

6.7.1

Qu. 7A *Of what do the walls of the radial artery consist?*

Feel for the pulsation of the ulnar artery as it crosses the flexor retinaculum.

Examine the pattern of superficial veins on the back of the hand and wrist, passing to the main cephalic and basilic veins. With the upper arm gripped firmly to prevent venous return, palpate the front of the elbow and feel for a vein into which hypodermic needles are often inserted. Here, the basilic and cephalic veins are linked by a cubital vein which also receives blood from deep tissues of the forearm. The ulnar artery may be anomalously superficial in this position.

Qu. 7B *How would you decide whether what you feel is artery or vein?*

Place your right index finger on the distal end of one superficial vein on the back of the forearm and occlude it. With the index finger of the other hand, express the blood from the proximal end of the vein. The vein is likely to remain empty. Harvey (1628) performed this simple experiment to illustrate that venous blood was returned to the heart through veins which contained valves.

Qu. 7C *What would be the explanation if the distal end of the vein was to fill with blood when you removed the pressure of your proximally placed left index finger?*

Using a red skin–pencil mark out on your partner the course of the major arteries to the upper limb.

B. Prosections

a. Arterial supply to the upper limb
(6.7.1)

Examine the prosected part and identify the **subclavian artery** as it arises from the arch of the aorta and crosses the first rib. As the artery enters the axilla between the clavicle and outer margin of the 1st rib it is renamed the **axillary artery**. In the axilla the artery is accompanied by the brachial plexus of nerves arising in the neck and thorax. In the natural state this neurovascular bundle is packed around and protected by fat. The axillary artery gains added protection in the axilla by virtue of the muscles of its anterior and posterior walls. The artery gives a branch (**acromio–thoracic artery**) which supplies the superficial tissues of the shoulder region anteriorly, and branches to the chest wall (**superior** and **lateral thoracic arteries**), which also supply the breast and become enlarged in lactation. It also gives off anterior and posterior **circumflex humeral** branches which encircle the upper end of the shaft of the humerus, and a **subscapular** branch which follows the lateral border of the scapula. As the artery leaves the axilla at the lower border of teres major it is named the **brachial artery**. This artery passes down the medial aspect of the arm to reach the front of the elbow. In this position it is relatively superficial and lies medial to the prominent tendon of the biceps, where its pulsations can be felt. From the brachial artery arise small anterior and poster-

ior arteries to supply the humerus and muscles, and the **profunda brachii artery** which accompanies the radial nerve in its spiral course around the humerus and supplies triceps and the elbow joint.

As the brachial artery passes into the forearm it divides into **radial** and **ulnar arteries** which continue towards the wrist lying between the superficial and deep muscles of the forearm. Soon after its formation the ulnar artery gives a branch which passes deeply into the forearm and gives rise to two interosseous arteries which run distally on either side of the interosseous membrane between the radius and ulna supplying the deep tissues. At the wrist the ulnar artery lies relatively superficially and passes into the palm lateral to the pisiform bone. At this point it divides into superficial and deep branches: the superficial branch turns laterally and, crossing the tendons of the forearm muscles which are passing into the fingers, anastomoses with a smaller superficial branch of the radial artery to form the **superficial palmar arch**. This lies at the level of the distal border of the outstretched thumb. The deep branch of the ulnar artery passes through the small muscles of the little finger (muscles of the hypothenar eminence) and, passing laterally beneath the long tendons, anastomoses with a deep branch of the radial artery, forming the **deep palmar arch** (see **6.7.1**, **6.12.2**, **6.12.3**). As the radial artery reaches the wrist its superficial branch crosses the small muscles of the thumb (muscles of the thenar eminence) to anastomose with the superficial branch of the ulnar artery. The continuation of the radial artery, however, crosses the lateral aspect of the wrist joint under cover of long tendons passing to the thumb (the anatomical snuff box) and, on the dorsum of the hand, passes between the adjacent heads of the 1st dorsal interosseous muscle, to lie in the palm deep to the long flexor tendons to the fingers; here, it anastomoses with the deep branch of the ulnar artery to form the deep palmar arch which lies at a level of the proximal border of the outstretched thumb. The fingers are supplied by **digital arteries** which arise from the superficial palmar arch, divide opposite the metacarpal heads, and pass along the sides of the fingers to their tips. The deep structures of the palm are supplied by branches of the deep arterial arch.

As you would expect, all the muscles of the upper limb are supplied by muscular branches which arise from adjacent arteries. However, it is worth remembering that, wherever a considerable amount of movement takes place (i.e. at joints or between parts of the body such as the scapula and the chest wall), the blood supply to the area will be plentiful and anastomoses between branches of nearby arteries will almost certainly be found.

Look for anastomoses of large arteries around the scapula or the elbow joint.

Find a nutrient artery which supplies the shaft of a long bone.

Qu. 7D *If someone had sustained a deep cut in the palm of the hand how would you stop the resulting haemorrhage?*

Qu. 7E *Why are the walls of arteries thicker and more elastic than those of veins?*

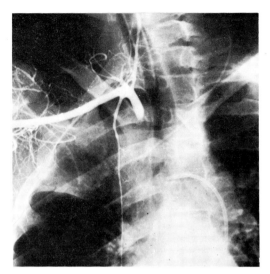

6.7.2

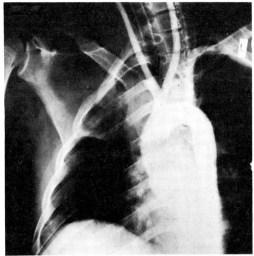

6.7.3

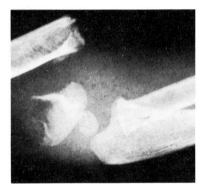

6.7.4

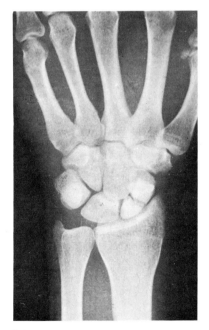

6.7.5

The following case illustrates how a knowledge of the arterial supply to the arm enabled a surgeon to trace and remove an acute blockage in a major vessel. A car driver went into a head–on collision with a lorry, he was severely injured and prior to operation it was observed that the right arm was abnormally cold. The pulse could not be felt either in the right wrist or in the axilla, as it could in the left arm. Arteriograms were taken to trace the site of the blockage to the circulation in his right arm.

Through a small incision near the elbow a plastic catheter was passed retrogradely (i.e. against the flow of blood) up the right brachial artery, to the axillary artery. Another catheter was passed retrogradely up the right femoral artery to the proximal end of the arch of the aorta. When the catheters were in place, radio–opaque medium was injected first into the axillary artery and radiographs taken (**6.7.2**). More opaque medium was then injected into the proximal end of the aortic arch and further radiographs were taken (**6.7.3**).

Qu. 7F *Examine Figs.* **6.7.2** & **6.7.3**, *where was the site of arterial obstruction in this patient?*

Ischaemic contracture (Volkmann's)
Supracondylar fractures of the humerus in children (**6.7.4**) occurring as a result of a heavy fall on the outstretched arm are relatively common. A complication of such a fracture is that of obstruction to the flow of blood along the brachial artery, due most commonly to reflex constriction of the vessel. Although gangrene of the hand or forearm does not usually occur, there may follow death of tissue deep in the forearm, the severity of which depends on the efficiency of the collateral circulation at the time. In severe cases the median and perhaps other peripheral nerves may be involved. As the name of the condition implies, the dead tissue contracts from fibrosis, and this leads to flexion contracture of the wrist, extension contracture of the meta-carpophalangeal joints, flexion contracture of the interphalangeal joints, and abduction and extension contracture of the thumb at its carpo–meta carpal joint. In addition there may be effects of a median nerve paralysis.

The blood supply of the scaphoid
In its course through the 'snuffbox,' lying lateral to the scaphoid, the radial artery gives off small branches to this bone which enter it at about the level of its 'waist'. Some branches pass to the proximal part of the bone, others to its distal aspect. A complete fracture across the waist of the scaphoid (**6.7.5**) may interrupt the blood supply to the proximal pole which then undergoes necrosis.

b. Venous drainage of the upper limb
(**6.7.6**)

The venous drainage of a limb can be considered in terms of:

a) The superficial drainage of the skin and underlying superficial fascia.
b) The deep drainage of structures which lie beneath the thick connective tissue (deep fascia) covering the muscles.

Digital veins lie on the medial and lateral aspects of the fingers and drain into a **dorsal venous arch** lying on the back of the hand. The medial (ulnar) aspect of this arch is drained by the **basilic vein** which passes upwards receiving tributaries from the medial aspect of the forearm. In the arm the vein pierces the deep fascia and is joined by deep veins running with the brachial artery which have drained the deep structures of the forearm and arm. Together they form the **axillary vein** which lies medial to the axillary artery in the axilla. The lateral (radial) aspect of the dorsal venous arch is drained by the **cephalic vein** which passes laterally up the forearm draining its superficial aspects. In the upper part of the arm it lies in the groove between pectoralis major and deltoid, a site at which venous catheters may be introduced. It then passes through the anterior wall of the apex of the axilla just beneath the clavicle to drain into the axillary vein. The pattern of minor veins is highly variable but there is usually a connecting vein, the **median cubital vein**, between the cephalic and basilic veins which lie in the front of the elbow. Since veins in the front of the elbow are anchored

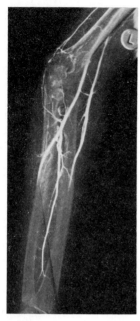

6.7.6

firmly by connective tissue to underlying structures, it is a convenient place to perform a venepuncture. Deep veins of the hand and forearm are usually found in pairs, **venae comitantes**, accompanying any small artery. These small veins eventually unite to form larger vessels which ultimately drain into the axillary vein. The axillary vein receives tributaries which correspond to the branches of the axillary artery, and become the subclavian vein at the outer border of the first rib. During exercise or in hot weather the superficial veins of the extremities dilate and aid heat loss by radiation and convection. Under resting conditions when the ambient temperature is cool, the characteristic arrangement of the deep vessels enables heat to be distributed by counter–current mechanisms from the small anteries to their accompanying venae comitantes, thereby helping to protect the core temperature.

Arteriograms

Examine the arteriograms of the upper limb (**6.7.6, 6.7.7**) and identify as many of the branches as possible. Note the anastomoses around the elbow and scapula. The arteriograms have been made by injecting a radio–opaque material into the circulation. It is a useful and informative technique in experienced hands but there are possible complications which preclude its use as a routine procedure.

A variation of arteriography is intravenous radionuclide angiography. A bolus of the patient's red cells, labelled with a suitable radioactive element such as Technetium–99m, is injected into the antecubital vein, and after passage through the heart and lungs it enters the systemic circulation. Using dynamic imaging techniques and computers it is possible to detect the path of the activity through most of the major vessels. While resolution comes nowhere near that of contrast arteriography it still gives a full functional answer with minimal risk to the patient. It is thus of particular value when dealing with children. A new and completely non–invasive technique for examining the vasculature is ultrasonic angiology which is used for investigating blood vessels of the thorax and abdomen more frequently than those of the limbs.

Lymphatic drainage of the upper limb
(**6.7.8**)

The lymphatic system is very difficult to demonstrate. **Lymph vessels** are not palpable in the living and, normally, **lymph nodes** are difficult to feel. The smaller vessels are difficult to dissect and, in the elderly, lymph nodes tend to be atrophic. The system is, however, extremely important in the spread and control of infection and malignancy.

Lymphograms (**6.7.9**) can be made by radiological examination of tissues after the injection of radio–opaque material into small lymphatic vessels and these are often used to detect larger vessels, lymph nodes (*arrow*) and sites of blockage of lymphatic drainage. Lymph nodes may also be

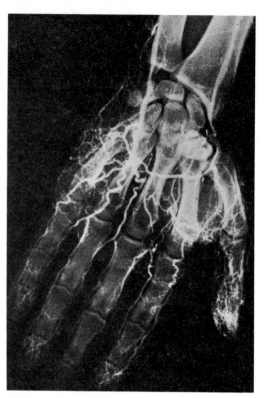

6.7.7

detected by the injection of radioactive colloids which are phagocytosed by the macrophage cells.

The lymphatic drainage of the upper limb parallels that of the venous drainage in that superficial lymphatics accompany the large superficial veins, and drain the surrounding skin and superficial fascia. They receive relatively few channels from the deeper tissues. Deep lymphatics follow the deep blood vessels and drain the deeper tissues. Thus, lymphatics from the thumb and thumb web, and the radial part of the forearm and arm, will tend to drain along vessels passing with the cephalic vein; those from the ulnar side of the forearm along the basilic vein. There is frequently a very small number of nodes situated just above the medial epicondyle (supratrochlear) which occasionally enlarge and can be felt.

Both superficial and deep vessels eventually drain into the axillary nodes. The first axillary nodes that they reach (lateral group) are situated along the axillary artery, and lymph passes then to nodes lying deeper (more centrally) in the axilla. The deepest (apical) nodes finally give rise to a **subclavian lymph trunk** which runs with the subclavian artery to end by joining the venous system at the confluence of the subclavian and internal jugular veins at the root of the neck. In addition to draining the upper limb, axillary nodes also drain the skin and superficial fascia of the trunk above the umbilicus, and the breast.

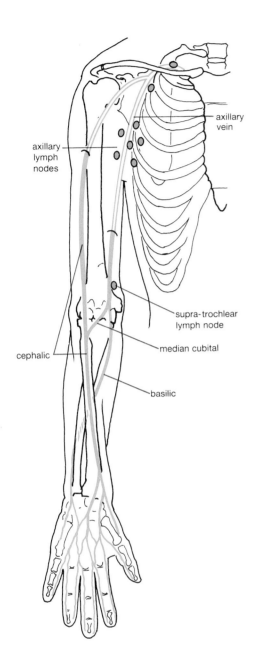

cephalic

axillary lymph nodes

axillary vein

supra-trochlear lymph node

median cubital

basilic

6.7.8

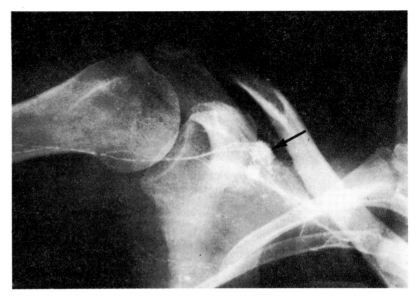

6.7.9

You should attempt to palpate lymph nodes in the axilla, remembering that the highest (apical) nodes are quite difficult to reach with the tip of the examining finger. A palpable lymph node is not necessarily an indication of active pathology, but may result from the fibrosis of a previously infected node. Lymph nodes draining infected areas are often enlarged and tender, and inflamed lymphatics can be seen as reddened cords along the path of the veins. Lymph nodes are also enlarged (but less frequently tender) when involved by neoplasms which may be benign or malignant, intrinsic to the node, or secondary to spread from a primary tumour in their area of drainage.

Qu. 7G *If you had a septic little finger, where would you be likely to find enlarged lymph nodes?*

Requirements:
  Prosections of arterial supply to the upper limb
  Prosections of major superficial and deep veins
    and their entry into the axillary vein
  Angiograms and lymphograms of the upper limb
  Skin marking pencils (red)
  Sphygmomanometer and stethoscope

Seminar 8

Innervation of upper limb: brachial plexus

The **aim** of Seminars 8–11 is to trace, by means of dissection, the course and distribution of the nerves to the upper limb.

A. Dissection

Blunt dissection, using your fingers, a pair of forceps, or by opening a pair of scissors in a space, is nearly always to be preferred to the use of a scalpel since it will enable tissues to be cleaved along anatomical planes. A scalpel should be used only when you are certain about what you are cutting. Make the relevant skin incisions illustrated in **6.8.1** and reflect the whole skin, i.e. the epidermis and dermis, from the underlying muscles by inserting your fingers or the handle of a pair of forceps between the dermis and the deep fascia overlying the muscles; do not use a scalpel unless it is essential. When your fingers are cleaving the plane between dermis and muscles, you will feel the resistance of certain cutaneous nerves which are passing through the subcutaneous tissue to supply the skin. Follow these nerves through the subcutaneous tissue and the dermis as far as possible before cutting them.

Having reflected the skin of the chest (including the mammary gland) and the upper arm, place your fingers deep to pectoralis major as it is nearing its insertion into the upper end of the humerus, and cut through it with a scalpel. Reflect the muscle towards the sternum. Reflect the pectoralis minor in the same way by cutting it close to its insertion at the tip of the coracoid process. The anterior wall of the axilla will now have been removed to expose the axillary neurovascular bundle which is surrounded by fat. Free the mid part of the clavicle from attached tissue and scrape away the periosteum. Next, cut away the middle of the clavicle using a bone saw and bone chisel, taking care not to damage structures lying deep to the clavicle. This will expose the neurovascular bundle as it crosses the 1st rib. Now remove the fat from the axillary bundle to reveal its contents.

Part of the 1st rib having been exposed, the lower attachment of a neck muscle, **scalenus anterior**, to its anterior aspect can be seen. This muscle, which can better be studied on a prosection of the neck, arises from the anterior aspect of the transverse processes of cervical vertebrae C3–6. The muscle is easily recognized since the subclavian vein lies in front of it as it crosses the first rib, while the subclavian artery and the nerve roots lie immediately posterior to it. Next, identify a pair of neck muscles, **scalenus medius** and **scalenus posterior**, which lie posterior to scalenus anterior.

They arise from the posterior aspect of the transverse processes of the cervical vertebrae and are inserted into the outer aspects of the 1st and 2nd ribs. From a consideration of the attachments of the scalene muscles it is clear that they must lie on either side of the intervertebral foraminae through which the cervical spinal nerves C5 to T1 emerge. (Remember that there are 8 cervical spinal nerves but only 7 cervical vertebrae.)

The formation of, and the branches between, the roots, trunks, and cords of the plexus are easy to identify and understand, but are difficult to retain in the memory for any length of time. In fact there is little to be gained from an intimate knowledge of this branching. But important aspects of the plexus, which you should remember, are:

- The relationships of the roots of the plexus in the intervertebral foramina.
- The position of the plexus on the side of the neck.
- The general motor distribution of the roots of the plexus.
- The roots involved in the reflex 'jerks' elicited during clinical examination.
- The general sensory distribution of the roots of the plexus.
- The major nerves arising from the plexus.

The relationships of the roots of the plexus in the intervertebral foramina are dealt with in the seminar on The Spine. Briefly, however, the **ventral and dorsal spinal nerve roots** emerge from the spinal cord within the spinal canal, and unite in an intervertebral foramen to form a **spinal nerve**. The **dorsal root ganglion**, containing cell bodies of the peripheral sensory fibres, is usually situated in the foramen. Immediately distal to the foramen the spinal nerve divides into **anterior and posterior primary rami**, the anterior rami forming the roots of the brachial plexus.

A knowledge of the position of the brachial plexus in the neck is particularly important for the anaesthetist who may be required to 'block' nerves or inject radio–opaque material into arteries in the vicinity. The 'posterior triangle' which lies on the side of the neck is bounded by the clavicle below, the posterior border of the sternomastoid anteriorly, and the anterior border of the trapezius posteriorly. It is called a triangle because the sternomastoid and trapezius muscles meet above, on the back of the skull. When the neck is extended and flexed laterally to the opposite side, and the shoulder depressed, the trunks of the brachial plexus can be felt as taut bands in the lower anterior angle of this triangle.

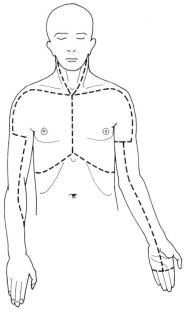

6.8.1

Between the scalene muscles in the cervical region, the spinal nerves form the **roots of the brachial plexus (6.8.2, 6.8.3)**. Identify these roots lying between scalenus anterior (anteriorly) and scalenus medius (posteriorly) and clean them of connective tissue.

From the roots, branches pass to supply the scalene muscles (segmentally); subclavius (C5 & 6); the rhomboid muscles via a nerve (C5) that passes back deeply through the muscles; and serratus anterior via the nerve to serratus anterior or **long thoracic nerve** (C5.6.7) which lies on the side of the chest wall and is vulnerable in operations for removal of the breast.

Qu. 8A *What action would be paralysed if the long thoracic nerve was damaged?*

Note that cervical roots C5 and C6 join to form the **upper trunk** (of the brachial plexus); the root of C7 remains single to form the **middle trunk**; and roots cervical C8 and thoracic T1 join to form the **lower trunk**. The trunks of the plexus lie relatively exposed on the side of the neck between sterno–mastoid and trapezius, where they can readily be felt.

From the upper trunk a nerve, the **suprascapular nerve** (C5, 6) passes back to supply supraspinatus and infraspinatus.

Qu. 8B *What actions would be paralysed if the suprascapular nerve was damaged?*

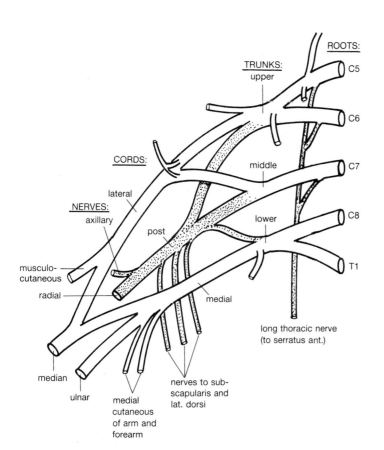

ROOTS:
C5
C6
C7
C8
T1

TRUNKS:
upper
middle
lower

CORDS:
lateral
post
medial

NERVES:
axillary
musculo-cutaneous
radial
median
ulnar
medial cutaneous of arm and forearm
nerves to sub-scapularis and lat. dorsi

long thoracic nerve
(to serratus ant.)

6.8.2

Occasionally the brachial plexus may be pre– or post–fixed so that the entire plexus may arise from a higher or lower set of segments. If the latter should occur, or if an extra (cervical) rib or fibrous band develops in the presence of a normally derived brachial plexus, then pressure may be exerted on the lower trunk of the brachial plexus. Since the lower trunk contains motor fibres which supply the small muscles of the hand, either through the median or ulnar nerves, damage to the root will cause wasting and weakness of the small muscles (see **6.9.5**). Sensory changes will also be found on the medial aspect of the forearm and hand. In addition, the autonomic fibres to the blood vessels of the upper limb may be irritated or damaged, and vascular changes occur which will be noticed predominantly in the hand.

Nerve bundles of the upper, middle, and lower trunks of the plexus become divided and redistributed beneath the clavicle such that **anterior divisions**, which supply flexor musculature, and **posterior divisions**, which supply extensor musculature, form the **cords of the brachial plexus**. There are three cords which are named according to their position in relation to the axillary artery. Identify the **lateral cord** and its branches: the musculo–cutaneous nerve which passes into the coraco–brachialis; the contribution to the median nerve; and the small branch to the pectoralis major (lateral pectoral nerve). Identify the **medial cord** and its branches which emerge between the axillary artery and vein: the contribution to the median nerve; the small medial cutaneous nerve of the arm; the larger medial cutaneous nerve of the forearm; and the ulnar nerve. Note that the medial cord also supplies both pectoralis major and pectoralis minor (medial pectoral nerve).

Qu. 8C *What is the function of a nerve plexus?*

Now tie off and cut the axillary vein as it crosses the 1st rib and remove it together with its tributaries which lie in the axilla. Pull the axillary artery laterally and identify the **posterior cord** of the brachial plexus and its branches; the axillary or circumflex nerve, which can be seen passing backwards through the axilla beneath the shoulder joint between subscapularis and teres major; and the radial nerve which runs downwards through the axilla to pass between the long and medial heads of triceps into the posterior compartment of the arm. Note also the branches of supply to subscapularis, teres major, and latissimus dorsi.

The motor distribution of the roots

Each muscle of the arm has a nerve supply which can be ascribed to more than one segment of the spinal cord. Remembering the root values of the nerve supply to individual muscles is unnecessary, and they can always be looked up in the reference textbook. It is more important to remember broad principles. Table 1 indicates in a general way which nerve roots are responsible for particular movements of the arm, since it is the movement of a joint that is more easily tested, rather than the action of

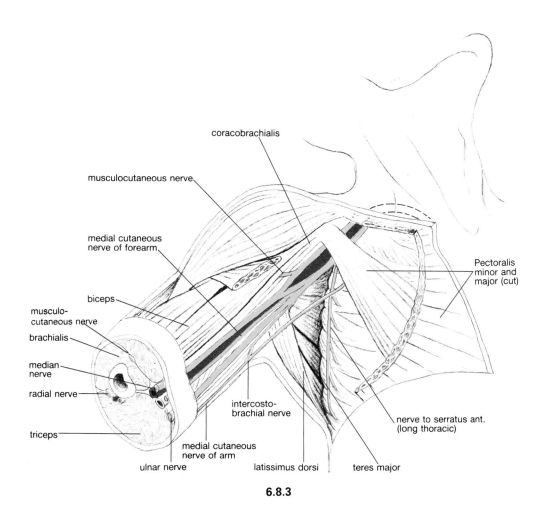

coracobrachialis

musculocutaneous nerve

medial cutaneous
nerve of forearm

biceps

musculo-
cutaneous nerve

brachialis

median
nerve

radial nerve

triceps

ulnar nerve

medial cutaneous
nerve of arm

intercosto-
brachial nerve

latissimus dorsi

teres major

nerve to serratus ant.
(long thoracic)

Pectoralis
minor and
major (cut)

6.8.3

Table 1. Segmental nerve supply to movements of the upper limb

Joint	Muscle action	Root supply	Muscle action	Root supply
Shoulder	Abduction Lateral rotation	C5	Adduction Medial rotation	C6, 7, 8
Elbow	Flexion	C5, 6	Extension	C 7, 8
Radio–ulnar	Supination	C 6	Pronation	C 7, 8
Wrist	Flexion and Extension	C6, 7		
Fingers	Long Flexors and Extensors	C7, 8		
Hand	Small muscles	T1		

particular muscles. By determining what movements of the arm are impaired by muscle paralysis a clinician can work out the level of neurological involvement of the plexus or spinal cord that may be present.

Each spinal nerve supplies a group of muscles derived from one myotome. There is a pattern of arrangement to the supply, but it is quite different from that of the dermatomes. In general, the more distal muscle groups are supplied by more caudal spinal nerves; opposite movements of a joint are supplied by adjacent spinal segments and the arrangement is based on movements rather than individual muscles.

The 'reflex jerks' of the arm elicited on clinical examination are stretch reflexes produced by sudden tapping of particular muscle tendons. The sudden contraction of a relaxed muscle after its tendon is tapped depends upon the integrity of nervous pathways from tendon and muscle receptors to the spinal cord, and also efferent motor pathways from the cord to the muscles concerned. The most common reflexes in the arm are elicited from the biceps tendon (biceps jerk) and the triceps tendon (triceps jerk) at the elbow. The root value of these reflexes can be found in the Table above, and loss of these normal reflexes implies impairment of the reflex pathway concerned.

The sensory distribution of the roots

The sensory distribution of the roots provides another method of determining the level of a neurological lesion. The pattern of loss of sensation in the arm may correspond to that of a major nerve, or of a larger part of the plexus. The area of skin that is supplied by a dorsal root of the spinal cord is known as a dermatome. Maps of the skin of the arm, as for the rest of the body, have been prepared on the bases of clinical cases of nerve root damage, and **6.8.4** is a diagram of such a map.

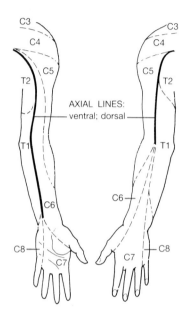

AXIAL LINES:
ventral; dorsal

6.8.4

The area of skin supplied by a spinal nerve is called a **dermatome**. As the arm grows out during development, skin innervated by the middle nerves of the group supplying the limb is taken distally.

It must be remembered that there is considerable overlap in the territories supplied by adjacent spinal nerves (and that different modalities have different areas of supply). This means that when one spinal nerve is lesioned, a neatly defined area of anaesthesia as suggested by **6.8.4** is *not* produced. As a general guide, you will see that the middle nerve of the group supplies the middle digits of the hand.

Between non–adjacent dermatomes (e.g. C5 and T1; C6 and C8) there is much less overlap. The lines between non–adjacent dermatomes are referred to as axial lines and are more pronounced on the flexor than the extensor surfaces of the limbs. This is made use of in the clinical testing of a patient for a 'sensory level' of a spinal nerve lesion.

By combining the findings on examination of the loss of sensation, of muscle paralysis and

reflex loss, the clinician can often work out, by a knowledge of the anatomy of the arm, a very accurate picture of the level of a neurological lesion. Most injuries to the brachial plexus occur as a result of severe traction, sometimes at birth. A fall from a motorcycle at speed is the commonest method of injury. The rider hurtles through the air, lands on the side of his head and on a shoulder, stretching the plexus, or else he strikes his shoulder on some obstacle such as a car, tree or lamp–post. A similar result can follow if a heavy weight falls onto the shoulder from above. These accidents tend to injure the upper root of the plexus, unless they are severe, when the middle and lower roots are involved as well, the nerves tearing in sequence from above downwards. The lower part of the brachial plexus can be injured in the opposite way, by sudden powerful traction of the arm above the head, as for instance when someone breaks a fall from the platform of a bus by holding onto the bar, or in falling through a trapdoor.

Suppose the **upper trunk** of the brachial plexus only is involved? Muscles which receive innervation from C5 and C6 will be paralysed and eventually waste. The most important are the clavicular head of pectoralis major, deltoid, supra– and infraspinatus, coracobrachialis, brachialis, biceps, brachioradialis, supinator and extensor carpi radialis longus. The arm hangs by the side, the elbow is straight, the arm is medially rotated and the forearm is pronated. This appearance has been called the 'waiter's tip' or 'crafty smoke' position.

Suppose the **middle trunk** (seventh cervical root) of the brachial plexus is involved? Usually it is damaged in addition to the upper roots and the following muscles are also paralysed: serratus anterior, latissimus dorsi, teres major, triceps, the middle fibres of pectoralis major, pronator teres, flexor carpi radialis, flexor digitorum superficialis, extensor carpi radialis longus and brevis, extensor digitorum and extensor digiti minimi. There is 'winging' of the scapula (**6.8.5**), loss of extension of the elbow, and radial deviation of the wrist and some weakness of power grip due to loss of extension of the wrist and fingers.

Qu. 8D *What other disabilities would you expect to be able to elicit?*

Suppose that the **lower trunk** of the brachial plexus only is injured? Then the muscles which receive their innervation from the 8th cervical and 1st thoracic segments of the cord will be paralysed. These comprise the lowest fibres of pectoralis major, all the muscles of the flexor compartment of the forearm except pronator teres and flexor carpi radialis, and the intrinsic muscles of the hand. The paralysis is much the same as would be produced if the median and ulnar nerves were divided. Muscles will waste, the grip of the hand is lost and a claw-like deformity, which will be described later, will result. With injury to the first thoracic nerve root the sympathetic supply to the head and neck is cut off on the side of the lesion. This will result in a 'Horner's syndrome' (see *Head and Neck,* Vol. 3). The skin of the little and ring fingers, and the ulnar side of the palm, wrist, and forearm is insensitive.

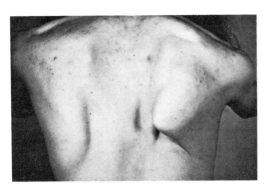

6.8.5

If the **whole brachial plexus** is injured the arm hangs uselessly from the shoulder, and is entirely insensitive except for the skin over the upper part of the shoulder supplied by the fibres from the supraclavicular branches of the cervical plexus (C4), and down the inner aspect of the upper arm supplied by the intercosto–brachial nerve (T2). Such an injury defies treatment. Nowadays, however, most paralysed arms are supported by splints which can be manoeuvred by the patient.

Qu. 8E *What other disabilities would you expect to be able to elicit?*

Do remember that it is not necessary for these lists of facts to be committed to memory. Table 1 (above) is provided so that you can work out which spinal nerve root supplies which muscle.

Requirements:
Articulated skeleton
Prosections of the neck and axilla showing the components of the brachial plexus: roots, trunks, divisions and branches.
Bone cutting tools

Seminar 9

Innervation of upper limb: lateral cord distribution

The **aim** of this seminar is to trace the course and distribution of the branches arising from the lateral cord (C5, 6) of the brachial plexus.

A. Dissection

a. The *musculocutaneous nerve* (6.9.1, 6.9.2)

The **musculocutaneous nerve** supplies the muscles of the anterior (flexor) compartment of the arm and the skin over the lateral aspect of the forearm. Trace the nerve as it leaves the axilla and pierces coracobrachialis. Find its muscular branches and then follow the nerve deep to biceps (if necessary divide biceps or coracobrachialis) until it emerges on its lateral aspect just above the elbow. Its terminal branch is the **lateral cutaneous nerve of the forearm** which supplies the skin down to the thenar eminence and which you should have located and followed, at least in part, when you reflected the skin or superficial fascia.

Qu. 9A *What disabilities would result from damage to the musculocutaneous nerve as it leaves the lateral cord?*

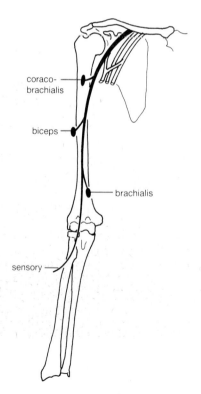

coraco-brachialis

biceps

brachialis

sensory

6.9.1

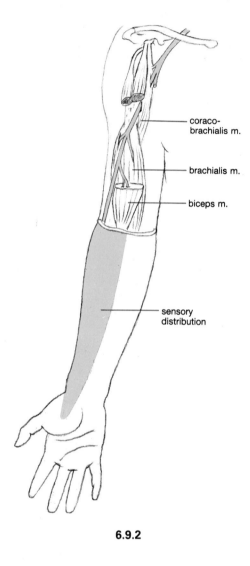

coraco-brachialis m.

brachialis m.

biceps m.

sensory distribution

6.9.2

b. The *median nerve* (6.9.3, 6.9.4)

This nerve receives contributions from both the lateral and medial cords. It supplies all the muscles of the anterior compartment of the forearm (with one exception: the flexor carpi ulnaris); the muscles of the thenar eminence and the lateral two lumbricals; and it supplies the skin over the palmar surface and the nail beds of the lateral (radial) $3\frac{1}{2}$ (or, sometimes $2\frac{1}{2}$) fingers.

Trace the nerve as it crosses in front of (or occasionally behind) the axillary artery to pass distally under cover of biceps through the forearm. At the elbow it lies medial to the brachial artery and therefore medial to the tendon of biceps. After it has crossed the anterior aspect of the elbow joint it passes into the forearm between the two heads of origin of pronator teres. As it does so it gives off a deep branch (anterior interosseous nerve) which passes with the interosseous branch of the ulnar artery to the deep muscles of the forearm. The median nerve is well protected in this part of the upper limb, lying between the superficial and deep muscles of the forearm; it passes distally, held closely to the deep aspect of flexor digitorum sublimis by fascia, until it reaches the wrist. Before it reaches the wrist, it gives off a slender **palmar cutaneous branch** which passes over the flexor retinaculum to reach the palmar skin.

As the median nerve crosses the wrist joint it lies between flexor carpi radialis and the lateral side of the long tendons of the superficial flexor muscles. The nerve is quite superficial at this point and could be easily infiltrated with local anaesthetic; however, this is only rarely performed since the nerve almost immediately enters the carpal tunnel where any subsequent inflammation may lead to pressure on the nerve and consequent dysfunction.

Before dissecting any further, look carefully at the prosection of the carpal tunnel and note the attachments of the thick band of fibrous tissue

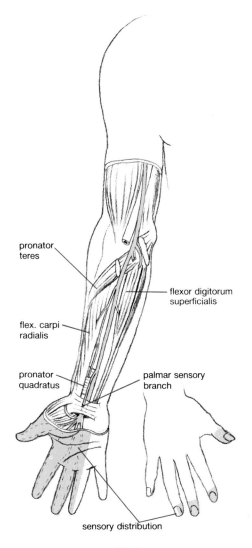

6.9.4

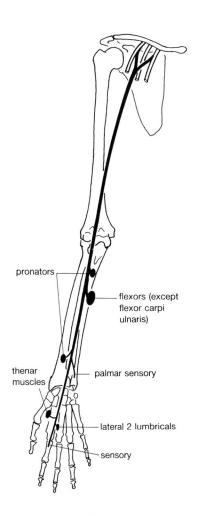

6.9.3

(flexor retinaculum) which is holding the long tendons to the fingers in place on the front of the wrist. It is attached to the pisiform bone and the hook of the hamate medially and the scaphoid and trapezium laterally. Trace the median nerve through this fibro–osseous tunnel and note the short, thick recurrent branch which leaves the nerve as it emerges from the tunnel to double back and supply the muscles of the thenar eminence. Lastly, identify the digital branches to the lateral $3\frac{1}{2}$ fingers and to the lateral two lumbricals; trace these branches into one finger noticing that they accompany the digital arteries which pass along the sides of the fingers. The digital nerves supply both palmar and dorsal aspects of skin over the terminal phalanx, and thus the nail bed.

Qu. 9B *If a finger needs to be anaesthetized where would you inject the local anaesthetic? Is there a possible danger in this procedure, and if so what is it?*

If the median nerve is damaged at the wrist, either by accidental injury or by compression in the carpal tunnel (carpal tunnel syndrome), there will follow:

• Paralysis of the thenar muscles (some muscles may not be affected and these are supplied wholly or in part by the ulnar nerve). On examination, the thenar muscles are wasted and the thenar eminence is flat with the thumb lying in the same plane as the fingers, an appearance often described as 'simian' since it is characteristically monkey–like (**6.9.5**).

• Paralysis of the lateral two lumbrical muscles, i.e. those to the index and middle fingers.

• Loss of sensation of the skin of the flexor aspects of the thumb, index and middle fingers, and the flexor aspect of the radial side of the ring finger (i.e. the radial $3\frac{1}{2}$ fingers), extending to the tips of the finger and over the nail beds. Loss of opposition of the thumb, and of sensation in the tips of the thumb and index finger prevent the normal pinch and feel between these digits which is of such importance in the normal discrimination of small objects and materials. Also, loss of opposition of the thumb interferes with normal grasp. This important movement can be restored if a tendon of the flexor digitorum superficialis muscle is re-routed and attached to the lateral border of the metacarpal of the thumb.

Qu. 9C *If the median nerve is damaged at the elbow or above what will be the additional effects to those mentioned above?*

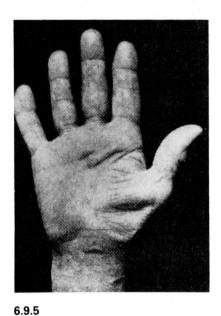

6.9.5

Requirements:
  Articulated skeleton
  Prosections of the axilla, arm and forearm showing the lateral cord of the brachial plexus and the distribution of the musculocutaneous and median nerves.

Seminar 10

Innervation of upper limb: medial cord distribution

The **aim** of this seminar is to trace the course and distribution of the branches arising from the medial cord of the brachial plexus.

A. Dissection

Note again the contribution of the medial cord to the median nerve and its supply to the pectoral muscles. Identify also the small medial cutaneous nerve of the arm which supplies the skin on the medial aspect of the arm; the larger medial cutaneous nerve of forearm and the ulnar nerve.

a. The *ulnar nerve* (6.10.1, 6.10.2, 6.10.3)

This nerve supplies the flexor carpi ulnaris; the ulnar half of the flexor digitorum profundus; the small muscles of the hand with the exception of

those which have been supplied by the median nerve, and the skin of the palmar and dorsal aspects of the medial 1½ (or 2½) fingers.

Trace the ulnar nerve, which is lying medial to the axillary artery in the axilla, distally through the arm to the posterior aspect of the medial epicondyle, where it can be palpated in the living body. Note that the nerve lies first in the anterior compartment of the arm and then pierces the medial intermuscular fibrous septum to enter the posterior compartment where it lies anterior to triceps. It then passes behind the medial epicondyle where it can easily be injured. If the nerve is hit or contused it may lead to a burning sensation which runs down the forearm to the little and ring fingers. Hence the layman's term of 'funny bone' for the bone on the medial side of the elbow! Now follow the nerve from behind the medial epicondyle distally, noting that it passes between the

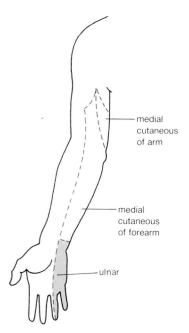

medial
cutaneous
of arm

medial
cutaneous
of forearm

ulnar

6.10.3

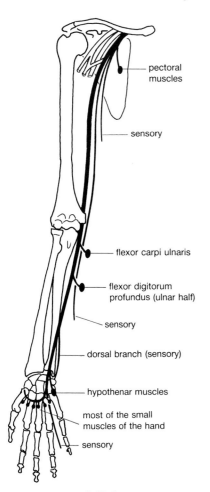

pectoral
muscles

sensory

flexor carpi ulnaris

flexor digitorum
profundus (ulnar half)

sensory

dorsal branch (sensory)

hypothenar muscles

most of the small
muscles of the hand

sensory

6.10.1

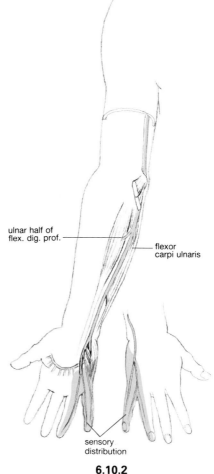

ulnar half of
flex. dig. prof.

flexor
carpi ulnaris

sensory
distribution

6.10.2

humeral and ulnar fibres of origin of flexor carpi ulnaris to reach the anterior compartment of the forearm. In the forearm, the ulnar nerve continues distally under flexor carpi ulnaris, thus lying on the deeply situated flexor digitorum profundus to which it gives a branch. As it nears the wrist a **dorsal cutaneous branch** is given off which passes onto the posterior aspect of the wrist to supply the dorsal aspect of the median $1\frac{1}{2}$ (or $2\frac{1}{2}$) fingers. As the ulnar nerve crosses the wrist, radial to the tendon of flexor carpi ulnaris, it lies superficial to the flexor retinaculum, to reach the small muscles of the little finger (hypothenar eminence muscles). The nerve then divides into superficial and deep branches. The **superficial branches** can be felt in the living body by feeling for the hook of the hamate on the hypothenar aspect of the palm. Roll the pad of your thumb gently over the prominent hook and you should be able to feel the cord–like superficial branch of the ulnar nerve as it passes across the bone. The superficial branches form digital nerves which pass on either side of the medial $1\frac{1}{2}$ fingers to the skin of the palmar aspects. The **deep branch** of the ulnar nerve passes first through the hypothenar eminence muscles and supplies them. Follow the deep branch carefully as it then turns laterally beneath the long flexor tendons to the fingers to lie within the deep palmar arterial arch; it supplies most of the small muscles of the hand. That is: the dorsal and palmar interossei, the medial two lumbricals, and the adductor pollicis. To trace the nerve satisfactorily, you should cut the long tendons to the fingers if you have not already done so, above the wrist; also cut through the flexor retinaculum. Now reflect the long tendons towards the fingers and this should enable you to trace the deep branches of the ulnar nerve with ease. Though it cannot be readily demonstrated by dissection, it is important to note that digital branches of the median and ulnar nerves to the fingers supply not only the palmar aspect of the fingers, but also the nail bed via branches that pass dorsally around the distal phalanx.

Qu. 10A *What would be the results of severance of the ulnar nerve at the wrist?*

The ulnar nerve can sometimes suffer gradual traction and compression as a result of an increase in the carrying angle at the elbow during growth. If the angle is increased (**6.10.4**) the deformity of the elbow which results is known as one of valgus, and the opposite, one of varus. The varus deformity is also sometimes known, for obvious reasons as the 'gunstock' deformity. Both deformities can arise, for instance, after union of badly set supracondylar fractures. The valgus deformity can also occur gradually if an injury causes impairment of growth to the lateral side of the growth cartilage of the humerus at the elbow; the medial side of the growth cartilage continues to form new bone normally, but the lateral side lags behind. The elbow gradually becomes valgus as growth continues, and as it does so the ulnar nerve behind the medial epicondyle becomes stretched. If the deformity is not corrected, the ulnar nerve may become permanently damaged, and the patient will lose the power of

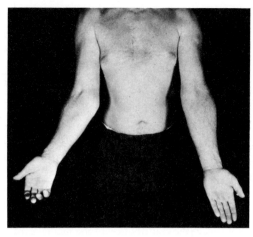

6.10.4

pinch between the thumb and index finger. A loss of power of abduction and adduction of the fingers can also be demonstrated by getting the patient to hold a piece of paper between extended fingers, or by trying to spread the fingers against resistance. There is also 'clawing' of the little and ring fingers, the metacarpophalangeal joints tending to hyperextend while the interphalangeal joints flex (**6.10.5**).

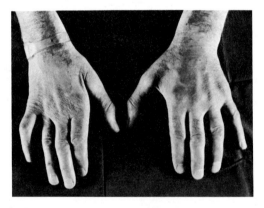

6.10.5

Qu. 10B *Why are the middle and index fingers not similarly clawed?*

The muscles of the hypothenar eminence waste, as do the interossei, producing characteristic grooves between the metacarpals when viewed on the dorsum of the hand.

Qu. 10C *If the ulnar nerve is damaged at the elbow or above what effects will be found in addition to those occurring after severance at the wrist?*

Requirements:
Articulated skeleton
Prosections of axilla and forearm showing the course and distribution of the medial cord of the brachial plexus and its branches: ulnar; medial cutaneous of arm and forearm; and the contribution to the median nerve

Seminar 11

Innervation of upper limb: posterior cord distribution

The **aim** of this seminar is to trace the course and distribution of the branches arising from the posterior cord of the brachial plexus.

A. Dissection

The posterior cord (**6.11.1**) lies immediately posterior to the axillary artery. The cord gives off three branches, one to each of the muscles of the posterior wall of the axilla: the upper nerve to subscapularis, which is the most superior of the three; the lower nerve to subscapularis (which also supplies teres major); and the nerve to latissimus dorsi.

Before identifying and tracing the two major branches of the posterior cord, identify the **nerve to serratus anterior (long thoracic nerve)** (**6.8.3**) which does not arise from the posterior cord, but from contributions from the 5th, 6th, and 7th cervical roots. The long thoracic nerve passes over the 1st rib, behind the trunks of the brachial plexus, to run vertically downwards on serratus anterior which it supplies.

The serratus anterior muscle sometimes undergoes apparently spontaneous paralysis. Such paralysis has, however, sometimes followed the carrying of a heavy rucksack for a prolonged period, which suggests that in such a case the nerve to serratus anterior might have been compressed by the clavicle as the nerve emerges on to the uppermost digitation of the muscle on the first rib. Paralysis of the serratus anterior results in 'winging' of the scapula when the arm of the affected side is held at length forwards, or else pushed against an obstruction (**6.8.5**). The vertebral border of the scapula, and especially the inferior angle becomes prominent, because the paralysed serratus anterior no longer holds the bone close to the chest wall. Such paralyses usually recover.

It is worth remembering that the muscular walls of the axilla are supplied by nerves arising from all three cords of the brachial plexus and the long thoracic nerve. The anterior wall (pectoralis major and minor) is supplied by the medial and lateral pectoral nerves from the medial and lateral cords. The posterior wall (subscapularis, teres major, and latissimus dorsi), is supplied by branches arising from the posterior cord. The medial wall (serratus anterior) is supplied by the long thoracic nerve.

a. The *axillary (circumflex) nerve* (**6.11.2**)

Trace the **axillary nerve** (which you have done before) backwards as it passes between the lower

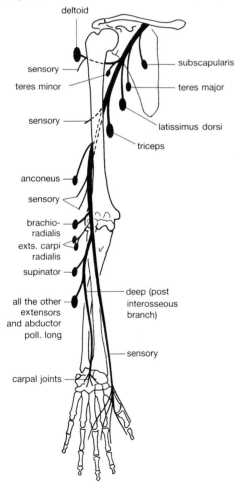

6.11.1

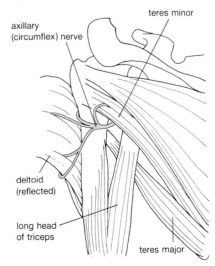

6.11.2

border of subcapularis and the upper border of teres major. As it does so, it supplies a branch to teres minor and then divides into two terminal branches. The deeper passes laterally around the 'surgical neck' of the humerus (where the bone is often fractured) and supplies deltoid from its deep aspect; the other, which lies parallel to the deep branch but passes superficial to deltoid, also supplies the muscle together with a small area of skin over the insertion of deltoid. Now turn the body on to its ventral surface. Define the posterior border of the deltoid and in so doing identify the superficial branch of the axillary nerve which is passing around this border of the muscle to reach the area of skin over its insertion. When tracing the axillary nerve through the posterior wall of the axilla you should have noticed its close proximity to the weakly supported inferior aspect of the shoulder joint. Pass your index finger along the course of the axillary nerve as it leaves the axilla and press upwards thereby palpating the lower aspect of the shoulder joint. You will remember that dislocation of this joint usually occurs in an inferior (and anterior) direction, so that the axillary nerve is sometimes injured.

b. The *radial nerve* (6.11.3, 6.11.4)

With few exceptions the **radial nerve** supplies the muscles and skin over the whole of the posterior aspect of the upper limb. Its branches tend to leave the parent trunk early.

Trace the nerve as it leaves the axilla and passes backwards into the posterior compartment of the arm. As the nerve passes between the long and medial head of triceps it gives off a small **posterior cutaneous nerve of the arm** which runs distally on the posterior aspect of the arm. As the radial nerve spirals laterally around the shaft of the humerus it lies superficial to the uppermost part of

the medial head of triceps; this provides some protection for the nerve in that the muscle fibres often protect the nerve from injury if the bone is fractured at this level.

Follow the spiral course of the nerve by cutting through the lateral head of triceps in an oblique direction; you will find that it lies close to the bone (in the 'spiral groove') with branches to the different heads of triceps and a small deep muscle, anconeus. As the nerve approaches the lateral intermuscular fibrous septum—which it pierces to enter the anterior compartment—it gives off a small cutaneous branch which supplies the lower lateral aspect of the arm (**lower lateral cutaneous nerve of the arm**) and a larger cutaneous branch, the **posterior cutaneous nerve of the forearm**, which passes distally down to the skin of the posterior aspect of the forearm. After the radial nerve has entered the anterior compartment of the arm it supplies the muscles arising from the lateral supracondylar area of the humerus (i.e. brachio–radialis and extensor carpi radialis longus). It lies under cover of these muscles on the lateral side of the arm and is thus well protected; follow it distally onto the anterior aspect of a muscle which wraps around the head of the radius, the supinator. At this point the radial nerve gives off a large and important branch, the **deep branch of the radial**, or **posterior interosseous nerve**, which supplies

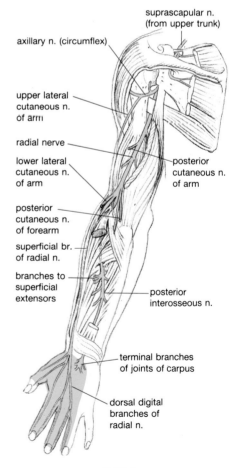

posterior cutaneous n. of arm

upper and lower lateral cutaneous n. of arm

posterior cutaneous n. of forearm

superficial branch of radial n.

6.11.3

suprascapular n. (from upper trunk)

axillary n. (circumflex)

upper lateral cutaneous n. of arm

radial nerve

lower lateral cutaneous n. of arm

posterior cutaneous n. of arm

posterior cutaneous n. of forearm

superficial br. of radial n.

branches to superficial extensors

posterior interosseous n.

terminal branches of joints of carpus

dorsal digital branches of radial n.

6.11.4

superficial extensors arising from the common extensor origin, pierces supinator and winds laterally around the neck of the radius to reach the posterior compartment of the forearm. Here, it supplies all the remaining forearm extensor muscles and supinator. Trace this branch into the posterior compartment of the forearm by dividing brachio–radialis and cutting through the fibres of supinator; identify at least one branch to an extensor muscle. Now return to the main body of the radial nerve as it lies anterior to supinator and follow it downwards through the anterior compartment of the forearm; you will find it lying between brachio–radialis and the deeper muscles of the forearm (in particular, flexor pollicis longus). As it approaches the wrist, the nerve passes dorsally deep to the tendon of brachio–radialis and then breaks up into superficial **dorsal digital branches** which supply the skin of the lateral $3\frac{1}{2}$ (or $2\frac{1}{2}$) digits on the back of the hand as far as the distal phalanx. It is possible to feel some of these branches by extending your thumb and palpating the skin over the tendon of extensor pollicis longus. The radial nerve is susceptible to injuries in the upper arm. If long crutches are used incorrectly the body weight may be taken by the upper bar of the crutch pressing hard into the armpit, instead of by the grip of the hand on a lower bar and the pressure of the upper bar upon the side of the chest. Excessive pressure into the axilla can produce a paralysis of the radial nerve and therefore of the extensor muscles of the forearm, with subsequent 'wrist drop'.

Qu. 11A *If a patient had sustained a fracture of the shaft of a humerus and you suspected damage to the radial nerve, what results would you expect?*

Requirements:
Articulated skeleton
Prosections of the axilla, arm, forearm and hand showing the posterior cord and its branches to the posterior wall of the axilla, the axillary nerve and the course and distribution of the radial nerve.

Seminar 12

The hand

The **aim** of this Seminar is to study the hand, and in so doing to revise the muscles of the forearm the blood and nerve supply to the hand, and its lymphatic drainage, all of which have been studied previously.

A. Living anatomy

Familiarize yourself again with the bones of the hand; with the superficial branch of the ulnar nerve which can be palpated as it is lying anterior to the hook of the hamate, and the superficial branches of the radial nerve crossing the tendon of extensor pollicis longus; with the sites at which pulsations of the radial, ulnar, and digital vessels can be felt, and the dorsal venous arch.

Remember that the functional axis of the hand passes through the centre of the palm and middle finger. Abduction and adduction of the fingers are related to this plane. The middle finger can therefore only be abducted from the midline. The attachments of the interossei reflect this functional axis.

B. Radiographs

Familiarize yourself again with the X–ray appearances of the hand in adult life and during development.

C. Prosections

a. The hand

Look at the form and distribution of the palmar aponeurosis. The palmar aponeurosis is clinically important because of the common condition which affects it, called Dupuytren's contracture (**6.12.1**). The palmar aponeurosis lies immediately deep to the skin of the palm, being separated from it by a layer of fat divided into multiple little loculi by fibrous tissue. It extends distally from the flexor retinaculum, dividing into four slips, one to each

of the fingers. Each slip of the fascia is attached to the fibrous sheath of the finger, and to the deep transverse metacarpal ligament on either side. Other fibres pass along the sides of the proximal phalanges to the base of the middle phalanges. In Dupuytren's contracture a dense fibrosis occurs in the aponeurosis, affecting especially the fibres to and of the little and ring fingers. Thickening of the fibrous tissue occurs and a flexion contracture of the metacarpo–phalangeal and proximal inter–phalangeal joints follows. Eventually severe deformity may result. The thickened, contracted aponeurosis needs to be excised by operation.

The fibro–osseous carpal tunnel at the wrist (see **6.6.14**) is firmly bounded by the flexor retinaculum and the concave aspect of the carpus. It is crammed with tendons to the fingers and thumb, and the median nerve (the ulnar nerve lies superficial to the flexor retinaculum). There is therefore no room for expansion within the carpal tunnel so that, should any swelling of the capsule of the carpal joints, or of the synovial coverings of the flexor tendons, occur, compression of the median nerve is liable to follow. Such changes can take place in pregnancy when water retention occurs, or in rheumatoid arthritis, and often for no very obvious reason! The condition is marked by tingling feelings in the thumb, index, and middle fingers, which may extend throughout the forearm and which are especially troublesome in the early hours of the morning. If the condition progresses the skin supplied by the median nerve becomes insensitive, and there is wasting and weakness of the thenar muscles. Persistent symptoms can be relieved by division of the flexor retinaculum at operation.

b. Blood supply to the hand (6.12.2, 6.12.3)

Examine again the superficial and deep palmar arterial arches which are formed by the anastomoses between the superficial or deep branches of the radial and ulnar arteries. Examine too the venous drainage of the hand; the dorsal venous arch and the cephalic and basilic veins which drain it. Remember also that the superficial lymphatic channels accompany the superficial veins while the deep lymphatics accompany the deep blood vessels.

c. Superficial aspect of the hand (6.12.6.)

i) Identify the tendons of the long flexor and extensor muscles as they cross the wrist to their insertions into the bones of the carpus and the

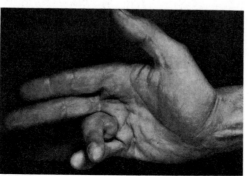

6.12.1

phalanges. Having identified them on the prosection, identify them on your own wrist (see **6.6.5**).

Examine the anterior and posterior aspects of your own wrist noting the positions of the tendons, nerves, and vessels. The prosection shows that the deep fascia—the extensor retinaculum—is thickened over the back of the wrist and the retinaculum is attached laterally to the lower end of the radius and medially to the pisiform and hamate bones; it helps to keep the tendons from 'starting up' or 'bow-stringing' when the muscles contract. The extensor tendons are each surrounded by a double layered synovial sheath in the region of the wrist joint, the function of which is to reduce friction around the tendon.

As the long flexor tendons pass towards the fingers, they too are kept from 'starting up' by the flexor retinaculum. They are surrounded by a common flexor synovial sheath which extends from about 3cm proximal to the wrist to the level of the heads of the metacarpal bones, except in the case of the tendon sheath to the little finger, which is continuous with the common flexor sheath and extends to the base of the distal phalanx and may communicate with the common flexor synovial sheath (**6.12.4**).

Qu. 12A *What might be the danger if a deep wound of the little finger became septic?*

ii) Examine the three superficial small muscles of the thenar eminence (**6.12.6**, **6.12.7**). **Abductor pollicis brevis** arises from the scaphoid and the flexor retinaculum and is inserted into the lateral side of the base of the proximal phalanx of the thumb; as its name suggests, it abducts the thumb. **Flexor pollicis brevis** is more medially placed and it arises from the trapezium and the flexor retinaculum to be inserted alongside the former muscle; it flexes the thumb across the palm. **Opponens pollicis** lies deep to the former two muscles and arises from the trapezium and the flexor retinaculum to be inserted into the whole length of the lateral border of the metacarpal of the thumb; it brings the pad of the thumb into contact (opposition) with the pad of the little finger. During exploratory movements, the muscles of the hand act in concert and the pad of the thumb can be opposed to that of any other digit. The three superficial muscles of the thenar eminence are supplied by the median nerve. Deep to them lies adductor pollicis.

Examine next the three small muscles of the hypothenar eminence (**6.12.6**, **6.12.7**) which correspond with those of the thenar eminence. **Abductor digiti minimi** arises from the pisiform bone and is inserted into the medial side of the base of the proximal phalanx; it abducts the little finger away from the median axis of the hand. **Flexor digiti minimi** lies lateral to the abductor and arises from the back of the hamate and the flexor retinaculum, to be inserted alongside the former muscle; it flexes the metacarpo–phalangeal joint of the little finger. **Opponens digit minimi** lies deep to the other two muscles, it also takes origin from the hook of the hamate and the flexor retinaculum, and is inserted into the whole length of the medial

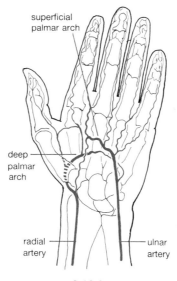

6.12.2

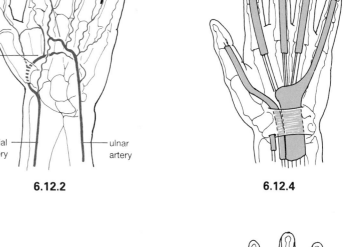

6.12.4

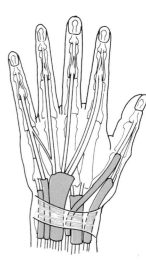

6.12.5

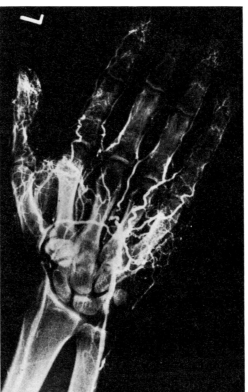

6.12.3

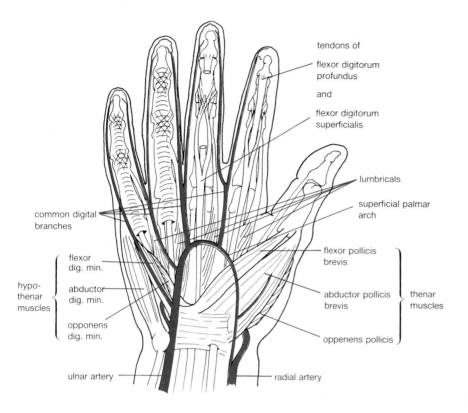

tendons of

flexor digitorum profundus

and

flexor digitorum superficialis

lumbricals

superficial palmar arch

common digital branches

flexor pollicis brevis

flexor dig. min.

hypo-
thenar
muscles

abductor dig. min.

abductor pollicis brevis

thenar muscles

opponens dig. min.

oppenens pollicis

ulnar artery

radial artery

6.12.6

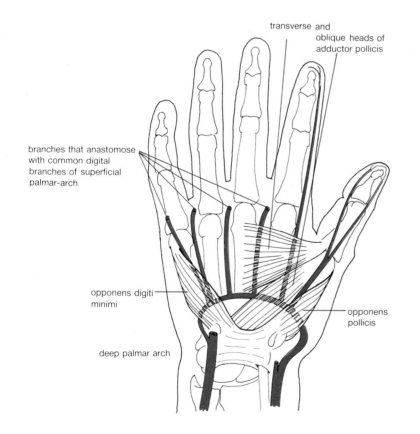

transverse and oblique heads of adductor pollicis

branches that anastomose with common digital branches of superficial palmar-arch

opponens digiti minimi

opponens pollicis

deep palmar arch

6.12.7

aspect of the shaft of the metacarpal bone; it opposes the pad of the little finger to that of the thumb.

The deep branch of the ulnar nerve sinks between the former two muscles to reach the deep aspect of the palm and, as it does so, supplies the hypothenar eminence muscles.

Abduct, flex, and oppose your thumb and little finger in turn against resistance, and feel for the contraction of the various muscles. These movements, especially that of opposition which is characteristic of Man, is essential for a 'precision grip'. Plastic repair of the hand after accidents involving the loss of the thumb entails fashioning a mobile stump which enables at least partial opposition to be restored.

d. Deep dissection of the hand (6.12.7)

Identify the four small **lumbrical** muscles which arise from the tendons of flexor digitorum profundus. They cross the radial side of the metacarpo–phalangeal joints of the fingers to be inserted into the dorsal extensor expansion. On contraction, they therefore flex the metacarpo–phalangeal joints and extend the interphalangeal joints. The two medial muscles are usually supplied by the ulnar nerve while the two lateral ones are supplied by the median nerve.

If the flexor retinaculum has been cut through, reflect the long tendons towards the finger tips and examine the **adductor pollicis**. You will find two heads of origin, one arising from the shaft of the middle (3rd) metacarpal, the other from the bases of the 2nd and 3rd metacarpals and the adjacent carpal bones. The muscle is inserted into the medial (and partly lateral) side of the base of the proximal phalanx of the thumb. It is supplied by the ulnar nerve, and its function is to bring the abducted thumb, which lies at right angles to the palm, back into contact with the index finger.

Examine the **interosseous** muscles. These are muscles which arise from adjacent sides of the shafts of the metacarpal bones and occupy the interosseous spaces between them. The **palmar interossei (6.12.8)** are small muscles which arise from the ulnar side of the 1st and 2nd metacarpals and from the radial side of the 4th and 5th metacarpals; they pass around the side of the proximal phalanx to which they are related and are inserted by tendons into the base of the dorsal extensor expansions. Their function is to adduct the fingers towards the midline of the hand which runs vertically through the middle finger; to flex the metacarpo–phalangeal joints and to extend the phalanges. The **dorsal interossei (6.12.9)** are larger and take origin by two heads, one from each of the metacarpal bones between which they lie. They insert both into the dorsal digital expansion and into the base of the proximal phalanges of digits 2, 3 (both sides), and 4. They act to abduct these fingers away from the mid–line of the hand, but share the other actions of the palmar interossei. The deep branch of the ulnar nerve supplies all the interosseous muscles.

Familiarize yourself once again with the course

and distribution of the nerves to the muscles and skin of the hand.

Nerve supply: The root value to all the small muscles of the hand is T1 (or the the lowest root of the brachial plexus).

The median nerve supplies the superficial muscles of the thenar eminence and the lateral two lumbricals; and the skin over the palmar aspect and nail bed of the lateral $3\frac{1}{2}$ digits.

The ulnar nerve supplies all the small muscles of the hand except for those supplied by the median nerve; and the skin over the dorsal and palmar aspects of the medial $1\frac{1}{2}$ digits.

The radial nerve supplies the skin over the dorsal aspect of the lateral $3\frac{1}{2}$ fingers and a small area over the base of the thenar eminence.

Note that the area of digital supply by all these nerves may vary by up to one digit.

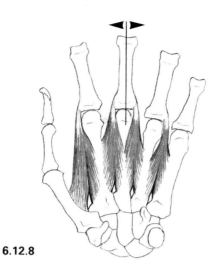

6.12.8

Rapid assessment of the major nerves of the arm

When confronted with a patient who has a badly injured elbow or forearm, it is important to assess not only the peripheral circulation of the limb but also the function of the major nerves. Often it is impossible, or too painful, for the patient to move the forearm, wrist, and fingers, so that reliance needs to be placed in the first instance on sensory function. A pinprick into the pulp of the terminal phalanx of the index finger will test the presence of median nerve function, and similarly of the little finger, the presence of ulnar nerve function. Pinprick over the dorsum of the first interosseous space will test the presence of radial nerve function, but this test is not quite so reliable. If extension of the thumb can take place at the interphalangeal joint, then the radial nerve must be intact to the level of the posterior interosseous nerve.

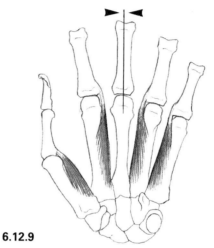

6.12.9

Requirements:
  Articulated skeleton
  Prosections of the hand showing *a*) superficial structures: superficial nerves and vessels, long flexor tendons and superficial palmar arch, muscles of the thenar and hypothenar eminences; *b*) deep structures: deep palmar arch and deep branch of ulnar nerve, adductor pollicis, interossei; *c*) flexor tendon sheaths.

Seminar 13

Skin and breast

The **aim** of this seminar is to study the skin of the upper limb and its embryological derivative on the chest wall, the mammary gland.

A. Living anatomy

Consider the structure of the skin in relation to its functions.

a. Protection

Examine the texture of the skin over the upper limb. It is coarser over the posterior aspect of the elbow and the whole of the extensor surface than it is over the flexor surface except over the palm of the hand.

Qu. 13A *Why is the skin thick over the palm of the hand?*

Qu. 13B *Why is it dangerous to burn your skin badly?*

Over the joints at which flexion and extension can occur (e.g. elbow) skin creases exist due to the presence of collagen fibres which attach the epidermis of the skin to the underlying deep fascia. Characteristic skin creases are also seen on the palm of the hand.

Qu. 13C *What is the underlying function of skin creases on the palm?*

b. Thermoregulation

Thermoregulation is achieved by a number of different mechanisms. Thyroid hormones control heat production by metabolism. The autonomic system controls heat loss in three ways: erection of hairs occurs in the cold but is of little importance in humans; the two major factors are the autonomic control of blood flow through peripheral cells which allow heat radiation, and the control of sweating.

Qu. 13D *Why does the skin of exposed areas go blue in cold weather?*

Using a hand lens locate the openings of eccrine sweat ducts on the forearm (see **3.1**). Eccrine glands are found over most of the body surface in Man and secrete a colourless watery saline; this secretion evaporates from the skin surface and assists in thermoregulation.

Qu. 13E *Can you make any comment with respect to the mineral content of sweat from the forearm?*

Examine the axilla and note the distribution of hair. The sweat glands in this region are largely of the apocrine variety. Their secretion occurs partly as a result of a breakdown and extrusion of the cells lining the luminal margins of the glands. If this secretion is not allowed to evaporate certain skin bacteria (staphylococci), by virtue of their enzymic actions, break down the secretion which results in odour formation. Some authorities have argued teleologically that this may act as a sex attractant; if this is correct then the axillary sweat secretion should perhaps be considered as a pheromone?

Examine the pads of your finger with the hand lens and notice the ridges and whorls of your finger prints; compare them with those of your partner. Watch your finger pad carefully and prick an adjacent finger pad with a pin. Unlike the eccrine sweat glands on the rest of the body you will note that these sweat glands secrete in response to 'alerting' or 'emotional' rather than thermal stimuli.

Qu. 13F *Do the sweat glands open on to the ridges or the furrows of the pad?*

Qu. 13G *What is the function of the finger pads with their superficial ridges and whorls?*

c. Cutaneous sensation

This has been considered in previous seminars.

B. Prosection of the mammary gland

After reading the sections which follow examine the prosection carefully.

The breast (6.13.1)

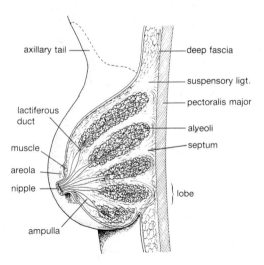

6.13.1

Examination of the living breast

When examining the breast of patients the usual procedure of visual and then manual examination is followed. The symmetrical contours of the breast in a number of positions of the arm should be noted, with the absence of any irregularities. The nipples are normally everted and the pigmentation of the areola varies, **6.13.2** shows a normal, non-lactating breast and should be compared with the lactating breast in **6.13.3**. A brown pigmentation usually remains after pregnancy in women who have borne children. Palpation of the breast for lumps is done initially with the palm of the hand rather than with the finger tips which tend to pick up too many small normal irregularities in the tissue. The extent of any abnormality should be noted and its attachment to other tissues determined by appropriate positioning of the patient or achieving contraction of pectoral muscles. Radiography, thermography (detection of infra-red emission from blood vessels), and scintigraphy using radionuclides are also used in the instrumental examination of the breast. In the mammogram **6.13.4** a breast carcinoma (arrow) distorts the local architecture and contains characteristic microcalcification.

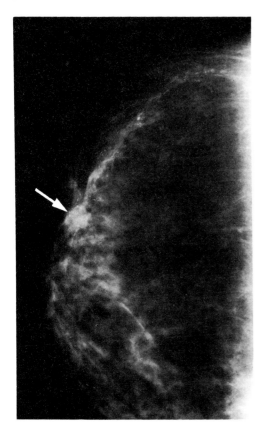

6.13.4

Development

The mammary gland is developed, as are the sweat and sebaceous glands, from columns of epidermal cells which proliferate and invaginate the mesoderm. Six or so invaginations occur along two vertical lines (mammary lines) which extend from the axilla, sweeping medially to the mid–clavicular line, and extending down to the groin. Normally, all but the pectoral pair of these invaginations (overlying the 3rd–5th ribs) disappear. Accessory breasts, or more commonly accessory nipples (the 'devil's marks' of medieval witch hunters), may be found at any point along the mammary lines but are more common in the pectoral region.

The pectoral pair of mammary primordia develop into compound ductal structures which arise from 12–15 main lactiferous ducts. These open on the surface of a raised nipple which is surrounded by an area of pigmented skin, the areola. In the immature female and the male, the mammary glands are similar: the breast tissue is rudimentary and confined to the margins of the areola. It comprises little more than a relatively unbranched ductal system in connective tissue stroma. The areolae are pink and the nipples relatively undeveloped. The structure of the gland remains rudimentary until about the time of puberty when, in the female, concentrations of plasma oestrogen increase and fat is laid down around the growing ductal system; the alveoli remain primitive. If pregnancy occurs, increasing plasma concentration of oestrogen and progesterone are responsible respectively for proliferation of the ductal system and the development of alveoli around the terminal ducts; the areola often becomes more deeply pigmented. At parturition, milk production is promoted by the secretion of prolactin from the anterior pituitary gland and a combination of various other hormones. However, the ejection of milk from the breast to the baby's mouth is caused by the neural stimulation of suckling which leads to an output of oxytocin from the posterior pituitary. This hormone causes the contraction of a network of specialized myoepithelial cells ('basket cells') which surround the alveoli, thus expressing milk into the ducts. After the menopause the mammary tissue tends to atrophy. Tissue in the male and immature female is rudimentary, but can respond to hormones. Maternal hormones may lead to the production of secretions ('witches milk') in newly born infants. Also, in the adult male, breast development may occur as a result of hormones produced abnormally in the body or administered for therapeutic purposes.

The mammary gland lies in the superficial fascia over pectoralis major and parts of serratus anterior and the external oblique muscle of the abdominal wall. Its form is highly variable, but the base is more constant, extending from the level of the 2nd to 6th ribs between the sternum and the anterior axillary line. The base of the normal breast should be mobile on the underlying chest wall. If it displays some tethering to the deep structures then this signifies disease. An axillary tail extends upwards and laterally into the axilla and frequently pierces the deep fascia. Less frequently, small pockets of breast tissue may pierce the deep fascia over the pectoralis major. Each breast consists of 15–20 lobes arranged radially around the nipple, and separated from each other by fibrous tissue septa. The lobes of normal glandular tissue cannot be palpated as discrete structures when the breast is examined with the palm of the hand. Each lobe is

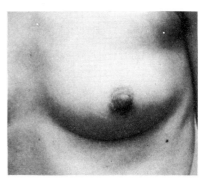

6.13.2

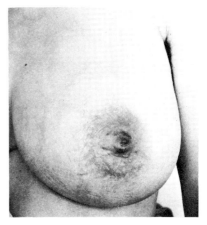

6.13.3

drained separately onto the nipple by its central lactiferous duct, which is dilated to form an ampulla just before it opens onto the skin. The nipple is surrounded by circular muscle fibres which erect it during suckling.

The areola has areolar glands (sebaceous) which secrete an oily lubricant and are most prominent in pregnancy and lactation. The virgin areola is pink but in pregnancy it takes on a brownish pigmentation which is never wholly lost.

The breast is supported by fibrous suspensory ligaments which extend from the deep fascia overlying pectoralis major to the dermis. Other fibres pass to the skin from the connective tissue of the breast. If the breast tissue is abnormally swollen with tissue fluid, these fibres invaginate the skin ('peau d'orange' appearance). In older women the fibrous supports atrophy and the breast tend to become pendulous.

The innervation of the breast is from the segmental intercostal nerves. The arterial supply and venous drainage come from two main sources. Medially, branches come from the internal thoracic artery and from intercostal arteries. Laterally, arteries reach the breast from the axillary artery (lateral thoracic and acromio–thoracic branches). The veins correspond to the arteries. It is noteworthy that the intercostal veins communicate directly with the vertebral veins and that secondary malignant deposits in the vertebrae are common.

The lymphatic drainage of the breast (**6.13.5**) is extensive and is of considerable significance if the tissue undergoes malignant change. Lymphatic vessels around the ducts drain the glandular substance of the breast, and those running in the fibrous processes between the lobules and lobes, drain both superficially to a subareolar plexus, and deeply to a submammary plexus in the deep fascia of the muscles covering the chest wall. Both from these plexuses and directly from vessels in the superficial fascia, lymph streams centrifugally from the breast to the regional nodes.

Probably most vessels pass through the fat of the axilla to the axillary nodes. Within the axilla lie some twenty to thirty nodes. However, some lymph drains to other groups of nodes, particularly the internal thoracic group within the chest. All of these need to receive attention in the treatment of breast cancer.

The axillary nodes are arranged into a number of overlapping groups: a pectoral group, behind the lateral border of the pectoralis major; a lateral, or brachial, group along the course of axillary vein, on its medial side; a subscapular group lying along the course of the subscapular artery; a central group lying on the floor of the axilla.

An apical group of a few nodes lies in the apex of the axilla. They give rise to a single subclavian trunk lying between subclavius and the subclavian

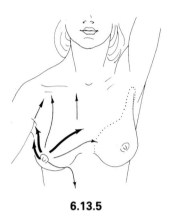

6.13.5

vein which empties into the angle between the subclavian and internal jugular veins, or into the thoracic duct on the left side.

These various groups intercommunicate, but in general the pectoral groups of nodes receive most of the lymph, passing it on directly to the apical group, or indirectly via the central group. The lateral group receives most of the lymph from the arm, and the subscapular group from the back of the upper part of the trunk and axillary tail of the breast, if it lies deep to the deep fascia.

Other groups of nodes receiving lymph from the breast include the internal thoracic group. There is a node in most of the upper six intercostal spaces alongside the internal thoracic artery, which drain the lymph from the inner half of the breast. They are very important since they are often involved at an early stage in the spread of cancer of the breast, moreover they communicate freely across the midline behind the sternum with the corresponding glands of the opposite side. Spread of cancer cells from these nodes to involve the subpleural lymphatic plexus can cause the escape of cancer cells into the pleural cavity, with secondary deposits and effusion of fluid into the cavity, and the rapid progress of the disease. One or two nodes may lie in the delto-pectoral groove alongside the cephalic vein, and others in the infraclavicular fossa. They receive some lymph from the upper part of the breast, and drain through the deep fascia to the apical axillary nodes. Some nodes also lie in the extraperitoneal tissue behind the upper abdominal wall. Lymph from the lower part of the breast drains through the abdominal wall, just below the xiphoid process of the sternum to vessels and a few nodes in the extraperitoneal fat. Cancer cells following this pathway pass through the diaphragm to mediastinal nodes. They may spread to the parietal peritoneum, and into the peritoneal cavity and to the surface of the liver. With the distribution of the regional lymph nodes in mind it is possible to devise a programme of treatment to combat the disease. If the surgeon elects to attempt a radical ablation of the disease, that is to remove the local lesion and as many of the regional nodes as is practical, his operation would have to include all the breast tissue, including the axillary tail, the deep fascia underlying it, the pectoralis major, the pectoralis minor and the interpectoral nodes, and all the fat and nodes of the axilla as high up as possible into the apex. If radiotherapy is to be undertaken after the operation this will be directed at the internal thoracic chain of both sides of the sternum, and at the supraclavicular nodes in the posterior triangle of the neck.

Requirements:

Prosections of the female mammary gland showing the extent of secretory lobules and their ducts opening onto the nipples.

CHAPTER 7

The lower limb
Introduction

Bipedalism in hominids means that the lower limb has become modified from the general pentadactyl pattern to function in both stance and the many varied forms of locomotion that are undertaken, often on very irregular surfaces. Unlike the arm, the lower limb frequently bears the entire weight of the body and this is regularly taken on just one leg in walking. Moreover, the stresses that the limb has to withstand are often multiplied many times, as on landing after a jump. In adapting to these functions the lower limb has foregone specialisations which enable the upper limb to perform precision movements over a wide range. Nevertheless, considerable dexterity can be attained with practise as, for example, in a person unable to use the upper limbs for this purpose.

The weight-bearing function of the lower limb demands that the joints are modified for stability in weight bearing. The weight of the body is transmitted through the shaft of each femur largely to the tibia in the lower leg. This bone, with the fibula, forms a mortice-type joint transmitting weight onto the talus, the bone which forms the apex of the arch of the foot. Finally, the arch transfers weight forward to the toes and backward to the heel.

The lower limb also requires considerable mobility in locomotion. To achieve this, the neck of the thigh bone has become elongated so that the shaft is offset from the head, an arrangement which also gives greater leverage to muscles acting at the upper end of the femur. Since the presence of the neck increases the distance between the upper ends of the femoral shafts, which are separated more widely than are the feet at ground level, the shafts slope downward and medially.

Furthermore, as a result of the medial rotation of the lower limbs during development not only are they brought closer together but the powerful extensor muscles of the thigh, leg and foot are positioned anteriorly, with the big toe lying medially. This is in contradistinction to the upper limb with its manipulative functions where the flexor muscles face anteriorly and the homologue of the big toe, the thumb, lies laterally. A major advantage which results from the different arrangement of muscles in the lower limb is propulsion where the combined strength of the extensor muscles can be exerted on the bones and joints to propel the body forward.

While studying the lower limb the anatomical specializations concerned with stance, which consists of weight bearing on one or both legs; with walking, which consists of a stance phase and a swing phase for each limb in turn; and with other forms of gait such as running and jumping, should be carefully considered.

Seminar 1

Bones of the lower limb

The **aim** of this seminar is to study the skeletal framework of the lower limb with respect to the living body and to the isolated bones and their x–ray appearances. However, this seminar is short and you will save time in the dissection of the nerves of the lower limb (Seminars 8 & 9) if you also start to remove the skin of the lower limbs on your cadaver (**7.7.1** for incision lines).

A. Living anatomy and the bony skeleton

Identify, on your partner, the skeleton and, on the radiographs, the bony points listed below (**7.1.2**, **7.1.3**).

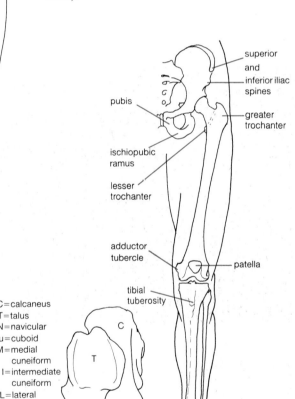

C=calcaneus
T=talus
N=navicular
Cu=cuboid
M=medial cuneiform
I=intermediate cuneiform
L=lateral cuneiform

7.1.1

7.1.2

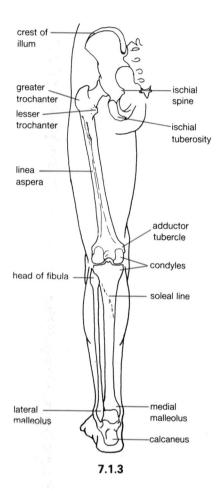

7.1.3

1. Pelvis

The bony pelvis has several important functions one of which is the protection and support of the pelvic viscera, and this will be studied in Volume 2. The pelvis also distributes the weight of the head and neck, upper limbs, and torso to the lower limbs and this aspect of its functions should be studied now. It comprises two innominate bones, each made up of an ilium, ischium, and pubis, joined together in the anterior midline by the pubic symphysis, and joined to the sacrum (Chapter 8) by the synovial sacro-iliac joints. Movements, other than those of a minor gliding variety at the sacroiliac joints, are prevented by strong ligaments and interlocking articular cartilage contours. Identify on the bony pelvis and yourself, where this is possible:

- the acetabulum, a laterally-facing cup-shaped hollow formed from the three pelvic bones; note the horseshoe-shaped facet in its roof with which the head of the femur articulates; the defect in its lower margin is spanned by a transverse ligament.

- the superior pubic ramus which extends laterally from the body of the pubis on each side above the obturator foramen.
- the ischiopubic ramus comprising the fused inferior ramus of the pubis and the ramus of the ischium which bounds the obturator foramen inferiorly.
- the ischial tuberosity, at the posterior extremity of the ischiopubic ramus, which takes the weight of the seated body and to which the sacrum is anchored by the sacrotuberous ligament thereby preventing backward tilting of the sacrum on the two innominate bones.
- the ischial spine to which the sacrum is attached by the sacrospinous ligament, the function of which is similar to that of the sacrotuberous ligament.
- the crest of the ilium and its tubercle.
- the anterior superior iliac spine at the anterior end of the iliac crest and the anterior inferior iliac spine a short distance below.
- the three buttresses of bone which resist weight-bearing stresses: (a) whilst standing on both legs, weight is distributed from the sacral articular facet to the acetabulum. (b) whilst standing on one leg, the line of maximum stress is directly vertically upward from the acetabulum to the tubercle of the crest. (c) while seated, weight is distributed from the sacral articular facet to the ischial tuberosity.

2. Femur

- the almost spherical head and its articulation with the horseshoe–shaped articular surface of the acetabulum; the pit in the head to which a ligament is attached (ligamentum teres)
- the neck of the femur and the angle it makes with the shaft
- the greater and lesser trochanters
- the anterior intertrochanteric line and posterior intertrochanteric crest
- the shaft with its posterior linea aspera which divides into the medial and lateral supracondylar lines at its lower end
- the flat popliteal surface of the femur which lies between the supracondylar lines
- the medial and lateral condyles and the shape of their surfaces which articulate separately with the tibia (indirectly via the incomplete intra–articular discs or menisci) and the patella
- the adductor tubercle.

3. Patella

Posterior articular surfaces for the condyles of the femur; their lower limit indicates the line of the knee joint cavity.

4. Tibia

- the upper surface articulating with the femur and articular menisci
- the medial and lateral condyles
- the tibial tuberosity for the attachment of the patellar tendon

- the shaft with its anterior (subcutaneous), medial, and interosseous borders; and medial, posterior, and lateral surfaces
- the laterally–placed facets for articulation with the head and lower end of the fibula
- the medial malleolus and its laterally–facing facet for articulation with the talus continuous with that on the lower end of the tibia.

5. Fibula

- the head with its apex and facet for articulation with the tibia
- the neck
- the shaft with its anterior, posterior and interosseous borders; and the lateral, posterior, and medial surfaces
- the inferior end of the bone with its facets for articulation with the tibia and the talus
- the lateral malleolus with its medially placed facet for articulation with the talus; and the malleolar fossa which gives origin to part of the lateral ligament of the ankle joint.

6. Bones of the foot

- the talus with its facets for articulation with the tibia and fibula, and with the calcaneus and navicular
- the calcaneus with its facets for articulation with the talus and cuboid; the posteriorly placed

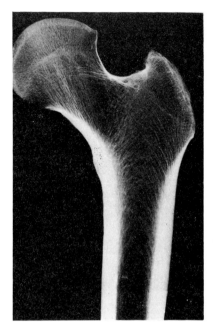

7.1.4

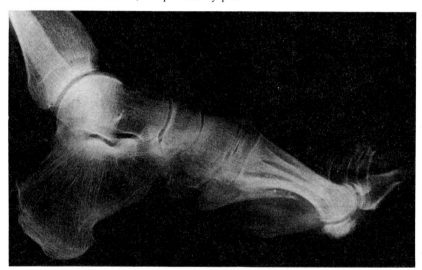

7.1.5

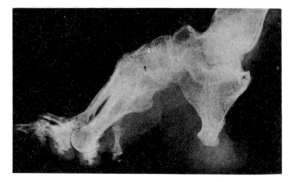

7.1.6

medial and lateral tubercles on its under surface; the large medial protuberance known as the sustentaculum tali
- the cuboid with its pronounced groove for the tendon of peroneus longus
- the navicular and its tuberosity to which the tendon of tibialis posterior is attached
- the medial, intermediate, and lateral cuneiform bones
- the metatarsals and the sesamoid bones at the distal end of the 1st metatarsal (the sesamoids are not usually retained in articulated skeletons)
- the phalanges.

The weight of the body is distributed between the posterior tubercles of the calcaneus and the heads and sesamoid bones of the metatarsals. The intervening bones forming an arch.

Qu. 1A   *Examine* 7.1.4, 7.1.5, 7.1.6 *and comment on the obvious trabecular patterns.* 7.1.6 *is of an aged woman from the orient whose feet were 'bound' from birth.*

Having identified the bony prominences mentioned in Section A it is important that you consider leg length. It is often necessary to measure the length of the legs of a patient as part of a clinical examination, and if a discrepancy between them is found, to determine where, in the skeleton of the leg, the difference lies.

With your partner lying supine on the couch, abduct one leg slightly, and then adduct the opposite one, so that the two legs again lie parallel to each other, though at an angle to the trunk. You will see that the abducted leg looks longer than the adducted leg. A measure of the difference between the 'apparent' lengths of the legs in this position can be made by firstly reading off the distance on a tape measure between the umbilicus and the medial malleolus of the abducted leg, and subtracting from it the similar measurement of the adducted leg; or simply by measuring the distance between the two malleoli. Do this and record your findings. Such 'apparent' shortening or lengthening of a leg may occur in a patient whose pelvis is tilted because of a fixed curvature of the spine, or because of an abduction or adduction deformity of one of the hip joints. The leg length difference is called 'apparent' because the difference is not due to any 'real' difference in the lengths of the legs, but to the position in which the legs lie in relation to the trunk.

The 'real' length of the leg can be measured from a fixed point on the pelvis. The anterior superior iliac spine is a convenient point. Record the measurement of your partner's legs from the anterior superior iliac spine to the medial malleolus on each side.

Are the lengths the same? Suppose there is an inch difference, how would you tell where the discrepancy lies? (it could, for instance lie in the leg below the knee)

Ask your supine partner to bend his knees to a right angle. If one femur is shorter than the other the difference becomes obvious and can be roughly measured.

Similarly if one shin is shorter than the other the difference is obvious, and can be confirmed by a measurement from the joint line of the knee to the medial malleolus. The tip of the greater trochanter, in the normal body, lies at the same horizontal level as the centre of the heads of the femur. Confirm this on a skeleton, and feel for your partner's greater trochanters. You can judge if they lie at the same level, or you can measure the distance between a finger tip placed on the greater trochanter and the highest point of the iliac crest. If there is a discrepancy between the levels of the greater trochanters then the true leg length difference lies in the region of the neck of the femur or hip joint. Perhaps the angle of the neck of the femur to the shaft is much reduced, or the joint is dislocated, or the neck of the femur is fractured! The length of the shaft of a femur can be measured from the tip of the greater trochanter to the line of the knee joint. The length of a femur is said to be about a quarter of the height of a body.

Record all the measurements you make.

Requirements:
Articulated skeleton
Separate bones of the pelvis, leg, and foot
Radiographs of the adult and juvenile bones of the lower limb.

Seminar 2

Joints of the lower limb

The **aim** of this seminar is to study the hip, knee, tibio-fibular, ankle, and subtalar joints, with the aid of the living subject and on prosected parts.

THE HIP JOINT

A. Living anatomy

Before examining the prosection of the hip joint look again at the articular surfaces on the skeleton and the x–rays. Move your own joint and note the range of movements which you can perform. These are similar to those at the glenohumeral joint since both these joints are synovial joints of the 'ball and socket' variety. The movements of the shoulder depend on movements not only of the glenohumeral joint but also of the scapula. Hold your partners right foot and ankle firmly and with both his legs fully extended rotate his leg medially and laterally. These movements are taking place at the hip. Test the degree of abduction and adduction which you find at the hip and record your findings.

Get your partner to lie on his side on the couch and pull his straightened leg backwards. Such extension as takes place is due largely to movement of the pelvis rather than the hip.

Now flex the hip until the thigh rests on the trunk. The normal hip joint appears to flex some 130° or so but, in fact, the hip joint itself can usually only flex about 90°–100°. Put your hand behind the lumbar spine of your partner as he lies flat on his back. Now flex the hip joint on the side on which you are standing. As the hip passes a right angle you will notice that the normal gentle forward curve of the lumbar spine (lordosis), flattens. Any flexion of the hip which occurs thereafter is caused by rotation of the pelvis in a transverse axis. Sometimes, in disease of a hip joint, the capsule contracts from scar tissue so that a flexion deformity develops. In other words the patient cannot extend the hip from the flexed to the anatomical position. Such a flexion contracture can be disguised, however, by increasing the lumbar lordosis, by rotating the pelvis enough to flatten the back of the thigh against the couch.

An understanding of how muscles produce medial and lateral rotation of the hip joint is complicated by the presence of the neck. The latter offsets the shaft and greater trochanter (to which the muscles are attached) from the centre of rotation which, in the standing position, passes vertically down through the centre of the femoral head to the lateral condyle. Medial and lateral

rotation occur with each step, as the pelvis swings. In the upright posture, muscles which run in front of the axis produce medial rotation, even though they are attached to the posterior surface of the femur (e.g. adductor longus); muscles which run behind the axis produce lateral rotation.

If however, the neck of the femur is broken, then the shaft of the femur is free to rotate on its own without constraint from the hip joint. In this case, powerful muscles, such as psoas major, produce lateral rotation of the femur. Patients with a fractured neck of femur therefore classically present with the uninjured leg in the normal position, and the injured leg in marked lateral rotation (**7.2.1**, **7.2.2**, **7.2.3**).

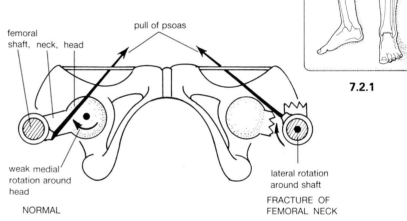

7.2.2

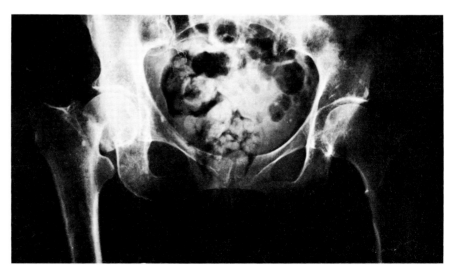

7.2.3

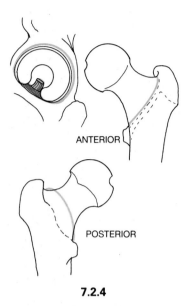

ANTERIOR

POSTERIOR

7.2.4

B. Prosections

Examine the capsule of the joint (**7.2.4**, **7.2.5**) and note its attachment to the margins of the acetabulum and the transverse ligament which bridges the gap in its inferior aspect. The distal attachment of the capsule is to the intertrochanteric line of the femur anteriorly, but posteriorly the capsule is attached around the middle of the neck of the femur. The capsule is reinforced by an **anterior ilio–femoral ligament**; the ligament is Y–shaped with its inverted stem attached to the anterior inferior iliac spine, and inferior attachments to the upper and lower parts of the intertrochanteric line of the femur respectively. This ligament is extremely strong and can act as a fulcrum when a dislocated head of femur is being repositioned in the acetabulum (reduction of a dislocated hip joint). The capsule is also reinforced by the **pubo–femoral ligament** which runs from the ilio– pubic eminence to the inferior aspect of the neck of the femur. Although an **ischio–femoral**

ligament is described as arising from the ischium posterior to the acetabulum, its fibres do not insert like the other two into the femur itself, but merge in a circular fashion into the capsule.

Examine the inside of the joint, noting the **horseshoe–shaped articular cartilage of the acetabulum**, the **pad of fat** which occupies the rest of the acetabulular surface and the **ligament of the head of the femur** (ligamentum teres). The latter arises from the transverse ligament of the acetabular rim to be attached to a pit in the head of the femur (it carries a blood vessel which provides nourishment only for the small area of the head to which the ligament is attached). As in the shoulder joint the depth of the acetabulum is increased by a ring of fibro–cartilage around its margin. The synovial membrane lines the capsule and non–articular structures within the joint such as the pad of fat and the ligamentum teres.

On the inner aspect of the distal part of the capsule you may see some fibrous bands or **retinaculae** which carry blood vessels to the formina of the neck and head of the femur (**7.2.6**). If these vessels are damaged due to a fracture of the neck, which is not uncommon in older people, the head may die and then crumble, destroying the joint.

Qu. 2A *With respect to stability and movement what are the major differences between the synovial 'ball and socket' joints of the upper and lower limbs?*

C. Radiographs

Examine **7.2.7** which is a radiograph of the pelvis and hip joints of an adult. Now examine **7.2.8** which is of the pelvis of a newborn baby. The os innominatum is seen to be made up of three separate centres, the ilium, the pubis, and the ischium. The acetabulum at this time is entirely cartilaginous, at the meeting place of the separate elements. The centre for the head of femur is not yet visible. **7.2.9** is a radiograph of the pelvis of an infant six months old; the centre of ossification of the head of the femur is now visible. Sometimes from the point of view of treatment, it is important to know whether or not the head of the femur is in the acetabulum at this time, since about one in a thousand babies are born with one or both hip joints liable to develop in a dislocated condition.

i) On the right side of **7.2.9** draw a line which passes across the pelvis horizontally from one acetabulum to the other, through the points where the separate elements of the os innominatum meet. Next draw a vertical line from the outer edge of the ilium, where it forms the outer lip of the acetabulum, to intersect the horizontal line. The centre of ossification of the head of the femur should lie in the lower inner quadrant.

ii) Also on the right side draw a curved line connecting the inferior aspect of the neck of the femur with the inferior margin of the superior pubic ramus. This line, known as Shenton's line, is smooth and continuous.

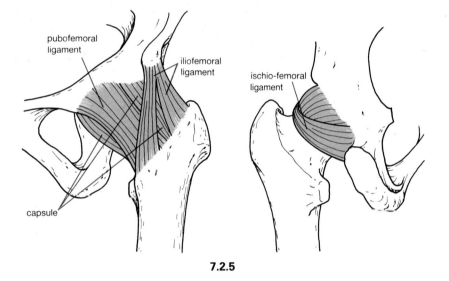

pubofemoral ligament

iliofemoral ligament

ischio-femoral ligament

capsule

7.2.5

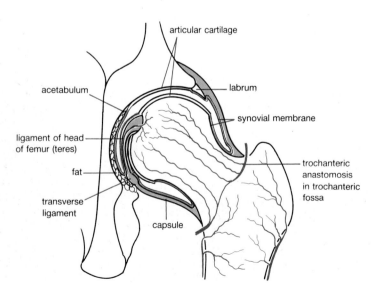

articular cartilage

acetabulum

labrum

synovial membrane

ligament of head of femur (teres)

fat

transverse ligament

trochanteric anastomosis in trochanteric fossa

capsule

7.2.6

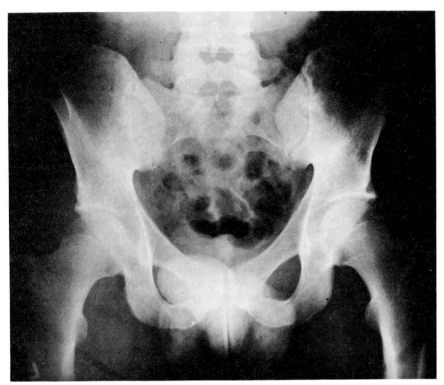

7.2.7

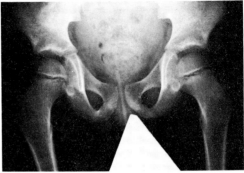

7.2.10

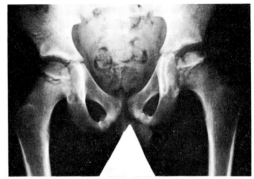

7.2.11

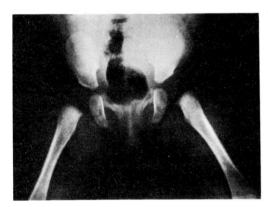

7.2.8

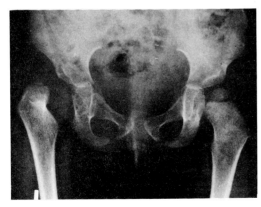

7.2.9

Qu. 2B *Trace the lines described on the left side of 7.2.9 and comment on your findings.*

7.2.10 is the radiograph of the pelvis of a child of seven. The centre of ossification of the head of the femur is now much larger, and forms a quite large 'cap' over the upper end of the femur, in the acetabulum. The acetabulum is more fully formed, and the centre of ossification of the greater trochanter is appearing. The blood supply to the epiphysis of the head of the femur is vulnerable between the ages of about five to ten years, depending almost entirely on vessels which reach the epiphysis along the upper and posterior aspect of the neck of the femur. If, for some reason, this blood supply is impaired, the capital epiphysis dies, in whole or in part, and the bone which is later reformed in the head of the femur may become flattened and misshapen; **7.2.11** shows the result of such a condition, known as Perthes' disease. The appearance of the diseased hip can be compared with the normal, opposite hip.

In a fifteen year old boy the head of the femur is well developed, and an obvious growth plate can be seen between the head and neck of the femur. In the adolescent great forces are transmitted through this cartilaginous growth plate, and it is not surprising that sometimes there is 'slip', either as an acute episode, or more usually as a subacute process, of the head of the femur on the neck (**7.2.12**). The head of the femur appears to slip inferiorly (as seen on an antero–posterior radiograph) and posteriorly (as seen on a lateral radiograph), so that the leg tends to take up a position of some external (lateral) rotation, with some limitation of abduction at the hip joint. This condition is known as 'slipped epiphysis', and is unique to the hip joint. Draw a line along the upper margin of the neck of the femur in the normal hip joint and

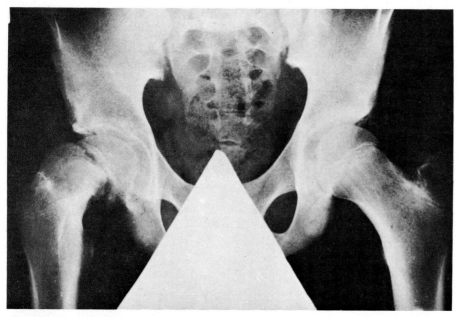

7.2.12

THE KNEE JOINT

A. Living anatomy

Look again at the articular surfaces of the knee joint on the skeleton and examine the radiographs **7.2.13, 7.2.14**.

Move your own joint; of what movements are you aware? (see later section on movements)

Define, on your own and your partner's knee joint the following parts of the joint:

- the full circumference of the patella, especially the medial border

- the articular margin of each femoral condyle

- the articular margin of the anterior parts of each tibial condyle

- the joint line anteriorly, on each side, and medially and laterally

- the medial and lateral epicondyles of the femur

- the tibial tubercle

- the adductor tubercle.

B. Prosection

Examine a prosected knee joint (**7.2.15, 7.2.16**) and note the extent of the hyaline articular cartilage which covers not only the condyles of the femur but also the lateral and medial tibial condyles and the posterior aspect of the patella. Note also the kidney and **semi–lunar shaped menisci** which are made of fibro–cartilage and lie between the lateral and medial condylar surfaces of the femur and tibia.

Since the patella articulates with the anterior surface of the lower end of the femur, the capsule must obviously be deficient on the anterior aspect of the joint. However, the muscle fibres of the quadriceps group insert into the patella superiorly while inferiorly the **patellar ligament** attaches the muscle to the tibial tuberosity. This arrangement, in addition to the fascial expansions (**retinaculae**) of the quadriceps muscle group which extend onto the tibial condyles, protects the joint anteriorly. Posteriorly the capsule is attached to the femoral shaft above the condyles and to the superior margins of the tibial condyles inferiorly. The **medial ligament** of the joint is broad and flat and extends from the medial epicondyle of the femur to the antero–medial aspect of the tibia below the medial tibial condyle; it is important to note that it is also attached to the medial meniscus. The **lateral ligament** is round and ropelike and is attached superiorly to the lateral epicondyle of the femur and inferiorly to the head of the fibula. The lateral ligament is not attached to the lateral meniscus but is separated from it by the **tendon of the popliteus**. This muscle, which is triangular in shape, arises from the back of the tibia above an oblique line which runs medially across the upper aspect of the shaft. It passes up through the capsule of the joint to attach to the lateral meniscus, and continues to reach a pit beneath the lateral epicondyle of the femur. The obliquely running **posterior ligament** spreads upwards and laterally, from the

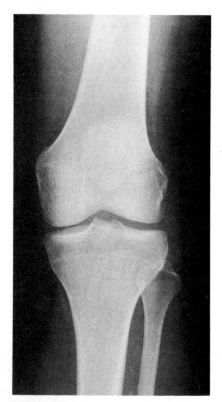

7.2.13

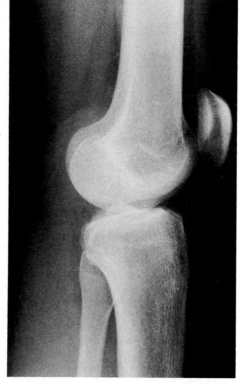

7.2.14

notice that is passes through the upper third of the capital epiphysis. This is one way of recognizing an early stage of slipping of the epiphysis in a young person complaining of pain in the hip, thigh, or knee.

Qu. 2C   *Draw this line on both sides of* **7.2.12**. *Which is the normal side and why?*

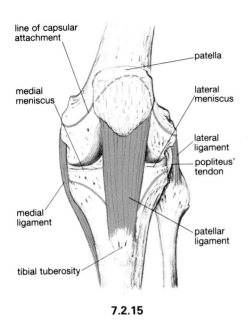

line of capsular attachment

patella

medial meniscus

lateral meniscus

lateral ligament

popliteus' tendon

medial ligament

patellar ligament

tibial tuberosity

7.2.15

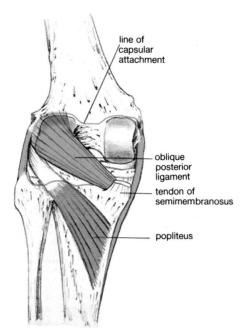

line of capsular attachment

oblique posterior ligament

tendon of semimembranosus

popliteus

7.2.16

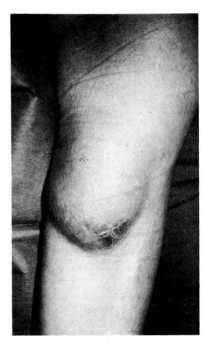

7.2.17

insertion of semimembranosus, reinforcing the capsule against torsional stresses. The **cruciate ligaments** are dealt with in the next section.

Intracapsular structures

The **synovial membrane** lines the capsule and those surfaces which are not covered by articular cartilage. Posteriorly, however, it covers the anterior but not the posterior surfaces of the cruciate ligaments and therefore does not line the whole of the posterior aspect of the capsule. Since the patella articulates with the femur, the capsule as well as the synovial membrane of the joint is deficient anteriorly; the synovial membrane therefore lines the patellar ligament and the retinaculae but it also extends above the upper pole of the patella, as the so–called **suprapatellar bursa**. It is obvious then that when a knee joint swells because of bleeding within it, or because of the accumulation of excess fluid as the result of inflammation of the joint, the swelling is a diffuse one which rises proximally, above the patella, as in **7.2.17**. This is in contrast to the sort of swelling that occurs because of the enlargement of the prepatellar bursa—'housemaid's knee'—which is confined to the front of the knee cap and does not communicate with the joint cavity. There are several bursae behind the knee joint. It is not necessary to remember them all, but one, commonly between the medial head of gastrocnemius and the capsule of the knee joint, can enlarge and present as a cyst behind the knee.

Examine the **intra–articular pad of fat** which lies below the patella, it is covered with synovial membrane and is attached by it to the inter–condylar notch of the femur. The fat pad has two wing–like expansions which are known as the **alar folds**. This arrangement not only increases the surface area of the synovial membrane but also helps to fill the changing spaces which occur during movement of the joint and, with the menisci helps to redistribute the synovial fluid.

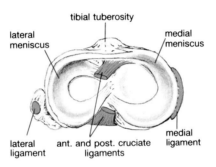

tibial tuberosity

lateral meniscus

medial meniscus

lateral ligament

ant. and post. cruciate ligaments

medial ligament

7.2.18

The fibro–cartilaginous **menisci** (**7.2.18**, **7.2.19**) are attached to the irregular central part of the tibia by each of their two horns. Peripherally they are attached by **coronary ligaments** to the capsule whence they gain a blood supply. This vascular supply is insufficient to provide nourishment for the inner parts of the cartilage which depend on the synovial fluid. When damaged menisci are removed, their peripheral parts may subsequently regrow. The medial meniscus is partly fixed by its attachment to the medial ligament while the lateral is attached not to the lateral ligament, but to popliteus which controls its movements. Severe strain (usually of a rotational nature) on a meniscus, can result in a longitudinal, or sometimes transverse, split in the fibrocartilage. This occurs more frequently to the medial disc since it is attached to the medial ligament of the knee joint and is presumably not quite as mobile as the lateral disc (**7.2.18**). The detached part may move into the centre of the joint and prevent the knee from extending fully—the 'locked' knee of sporting injury. **7.2.19** is an arthrogram of the knee joint; air has been introduced into the joint to show up the meniscus.

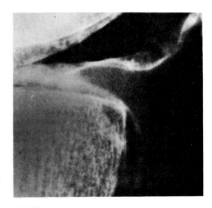

7.2.19

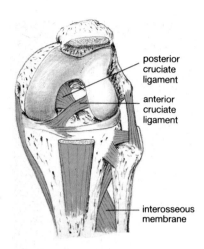

7.2.20 Anterior aspect of flexed knee joint.

The **cruciate ligaments** (7.2.20) are two thick rounded cords; the anterior ligament is attached to the anterior aspect of the upper surface of the tibia and passes upwards and backwards to the inner aspect of the lateral condyle of the femur. The posterior ligament is attached to the back of the upper surface of the tibia and runs upwards and forwards to the inner aspect of the medial condyle of the femur. These ligaments play a considerable part in the stability of the joint. If they are ruptured the tibia can be moved backwards and forwards with respect to the femur, a movement which is not possible in the normal joint; try this out on your partner.

Movements of the knee joint

You will have found that your joint can be flexed and extended, it therefore appears to be a straight-forward synovial joint of the hinge variety. However, more subtle movements take place at the joint which you cannot easily observe in the living, it is therefore termed a compound hinge joint. Take a separate femur and tibia and articulate them and also an articulated, dissected knee joint. Note that in the process of full extension the femur undergoes a small degree of medial rotation on the tibia due to the shape of its articular facets. In the process of extension the cruciate ligaments become taut but as medial rotation increases the ligaments become realigned and further extension, to the point at which the joint is 'locked', takes place. It is currently thought that, other than as a result of direct and violent trauma, the fully extended knee joint cannot be flexed unless popliteus contracts. Contraction of this muscle causes the femur to rotate laterally on the tibia thereby 'unlocking' the joint. Occasionally when a flexed knee joint is violently rotated while weight–bearing (as when kicking a football) a meniscus may be split.

Qu. 2D *If a footballer sustained an injury to a knee, and on examination it was found that the flexed lower leg could be moved backwards and forwards on the femur more easily than on the uninjured side, what would be the most likely cause?*

Qu. 2E *If, during the same game of football, another player complained of pain in the left knee which he was unable to straighten, what structure might he have damaged?*

THE TIBIO–FIBULAR JOINTS

The head of the fibula articulates with the postero–lateral aspect of the lateral condyle of the tibia; this superior tibio–fibular joint is a synovial one of the plane (or gliding) variety and the capsule is reinforced by strong ligaments. The shaft of the tibia is united for most of its length to the fibula by means of a fibrous sheet (interosseous membrane). Between the lower ends of the tibia and fibula, which are shaped to accommodate the wedge–shaped talus, there is a fibrous union or joint at

which very little movement can take place. Anterior and posterior ligaments and a deep transverse tibio–fibular ligament reinforce the inferior tibio–fibular joint and contribute to the socket into which the superior articular surface of the talus fits.

THE ANKLE JOINT

A. Living anatomy

Examine the articulated ankle and notice that the superior articular surface of the talus, which forms part of the ankle joint, is wider anteriorly than posteriorly. In dorsiflexion, therefore, as in walking up hills, the ankle is more stable than in plantarflexion, as in walking down hills. Now move your foot and discover how many types of movement you can make. Having examined the skeleton you will realize that the ankle joint between the lower ends of the tibia and fibula and the talus will allow of little movement other than extension (dorsiflexion) or flexion (plantarflexion). The movements of inversion or eversion of the foot take place only at the subtalar and transverse tarsal joints (see seminar 6).

B. Prosection

The capsule of the ankle joint (7.2.21) is attached around the articular margins of the tibia and fibula and the talus. As in other hinge joints, the capsule is reinforced by lateral and medial ligaments. The **lateral ligament** (7.2.22) is in three parts, each of which may be separately damaged in a 'sprain' of the joint. The **anterior talo–fibular part** is attached superiorly to the lateral malleolus and runs forwards and medially to the neck of the talus; the **calcaneo–fibular part** runs downwards and backwards from the tip of the malleolus to the lateral side of the calcaneum; the **posterior talo–fibular part** (7.2.23) runs horizontally from the malleolar fossa on the inner aspect of the malleolus to the posterior aspect of the talus. The fan–shaped **medial deltoid ligament** (7.2.24) is attached above to the medial melleolus of the tibia and below to the navicular and the sustentaculum tali of the calcaneus. It is very strong in comparison to the lateral ligament. When the talus is twisted in its mortice to the extent that ligaments or bones give way, it is more common (such is the strength of the medial ligament) that the medial malleolus is pulled off the shaft of the tibia, than that the ligament ruptures. Such a fracture is shown in 7.2.25.

Examine antero-posterior and lateral radio-graphs of the ankle joint (7.2.26, 7.2.27). Note the following points:

• in the adult there is overlap between the tibia and fibula at the inferior tibio–fibular joint. In the normal joint it is not possible to see between them. If you can, then a fracture of the joint has occurred.

• the ankle joint space is uniform.

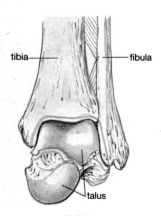

tibia — — fibula

talus

7.2.21

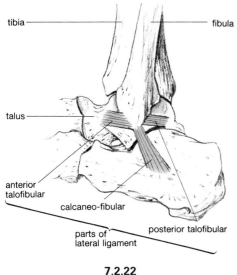

7.2.22

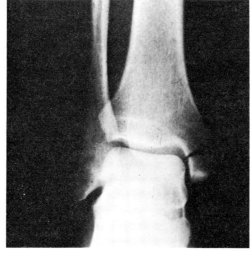

7.2.25

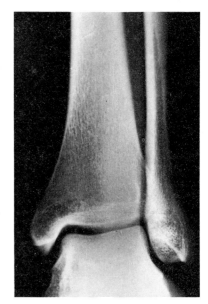

7.2.26

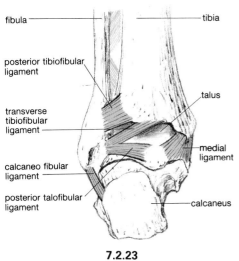

7.2.23

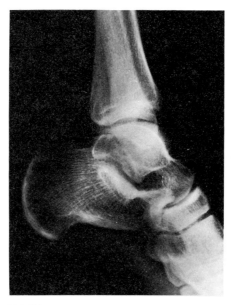

7.2.27

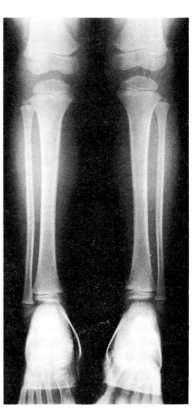

7.2.28

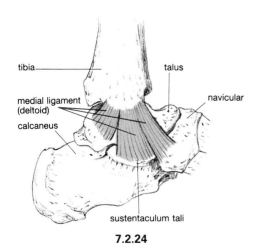

7.2.24

The maximum distance between the medial surface of the talus and the lateral, articular surface of the medial malleolus should be about the same. If it is much greater, then the talus has been shifted laterally and the medial ligament has been ruptured.

● in the young patient (**7.2.28**), the growth plate of the lower end of the fibula lies at the same level as the line of the ankle joint, at a lower level than that of the tibia. Note the plaster casts on both feet.

Qu. 2F *If a patient complained of pain in the left ankle after she had 'twisted' it by falling off a kerb, what structures might she have damaged?*

After you have completed the next three seminars make a list of the groups of muscles which act on the hip, knee, and ankle joint.

THE SUBTALAR JOINT

As its name implies, the subtalar joint (**7.2.29**) lies between the talus and the more distal bones of the tarsus. It consists of two parts, the articular surfaces of which are so arranged that they permit the movements of inversion and eversion of the foot (which are augmented by movements of the ankle joint). Look carefully at the lateral radiograph of the ankle and subtalar joints (**7.2.27**).

The posterior articulation (to which the term subtalar joint is sometimes restricted) is between the concave posterior facet on the under surface of the talus and the posterior convex facet on the upper surface of the calcaneus. A fibrous capsule is strengthened by **medial and lateral ligaments** and an **interosseous ligament** which occupies the bony canal between talus and calcaneus. The anterior articulation (sometimes referred to as the talo–calcaneo–navicular joint) is a ball and socket joint between the head of the talus and a socket

formed by the concave posterior surface of the navicular and the anterior facet on the upper surface of the calcaneus, and the upper surface of the 'spring' ligament. This **'spring' (plantar calcaneonavicular) ligament** (**7.2.30**) is a broad, thick band which connects the anterior margin of the sustentaculum tali to the navicular. It lies at the apex of the medial arch of the foot. The calcaneum and navicular are further connected on the dorsum of the foot by a **bifurcated ligament** which, in addition, connects the calcaneum to the cuboid. The **calcaneocuboid joint**, the apex of the 'lateral arch' of the foot, has a capsule strengthened dorsally by the bifurcated ligament and, on its plantar aspect, by the **short plantar ligament** and **long plantar ligament**. The short plantar ligament fills the depression between the anterior tubercle on the calcaneum and the ridge on the cuboid. The long plantar ligament extends from a broad attachment on the under surface of the calcaneum to the ridge on the cuboid and extends forwards to the bases of the 2nd to 5th metatarsals.

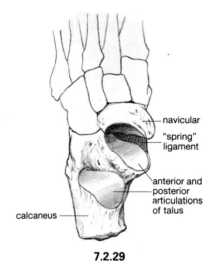

navicular
"spring" ligament
anterior and posterior articulations of talus
calcaneus

7.2.29

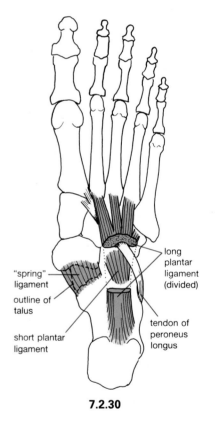

long plantar ligament (divided)
"spring" ligament
outline of talus
short plantar ligament
tendon of peroneus longus

7.2.30

Seminar 3

Gluteal region

The **aim** of this seminar is to study the gluteal region in the living and with prosected specimens.

A. Living anatomy

The iliac crest demarcates the gluteal region from the abdomen. Run your fingers backwards from the anterior superior iliac spine to the posterior superior iliac spine which underlies a dimple on the skin of the back opposite the 2nd sacral spine. The mass of the buttock is composed of muscle (gluteus maximus) and is limited below by the crescentic shaped fold of the buttock. Sit down on the palms of your hands and feel the bones on which you sit, these are the ischial tuberosities. Now, keeping your hands in situ, stand up. Repeat this movement and note how, when you sit down, the underlying gluteus maximus appears to rise and uncover the ischial tuberosities.

With your hands over gluteus maximus, extend your lower limb against resistance; at what degree from the vertical does this muscle contract?

The cutaneous nerve supply of the buttock is derived from branches of both lumbar and sacral nerves.

Investigate the actions of gluteus medius and minimus, and tensor fascia latae before you examine the prosections.

B. Prosections

Gluteal region

The **gluteus maximus (7.3.1)** is coarsely fasciculated and arises in a continuous line from the posterior part of the dorsal ilium, the lumbar fascia and the sides of the sacrum and coccyx; it also arises from the sacro–tuberous ligament which is attached to the side of the sacrum and the sacral aspect of the ischial tuberosity. The gluteus maximus is quadrilateral in shape and passes obliquely downwards and laterally to be inserted into the roughened area of the back of the femur below the lesser trochanter and into the ilio–tibial tract; the latter is a thick fibrous band which runs from the level of the greater trochanter to the lateral condyle of the tibia and into which the tensor fasciae latae (a muscle arising from the iliac crest) is inserted. Examine the innervation and blood supply of the gluteus maximus which comes from the inferior gluteal nerve and the inferior gluteal vessels. This neurovascular bundle passes out of the pelvis inferior to the piriformis. Gluteus maximus also receives a blood supply (but not innervation) from the superior gluteal neurovascular bundle.

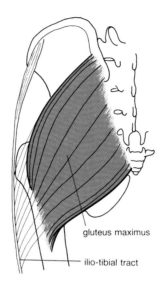

gluteus maximus

ilio-tibial tract

7.3.1

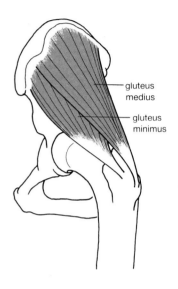

gluteus medius

gluteus minimus

7.3.2

Qu. 3A *When studying the living body you will already have noted that the gluteus maximus extends the trunk on the thigh, or conversely the thigh on the trunk, and is a prime mover in getting you out of your chair! What other action has this muscle?*

Reflect gluteus maximus to uncover **gluteus medius (7.3.2)** which arises from a large area of the dorsal surface of the ilium to insert into the lateral aspect of the greater trochanter of the femur; and **gluteus minimus** which arises from the dorsal ilium deep to gluteus medius and inserts into the anterior aspect of the greater trochanter. Both these muscles are supplied by the **superior gluteal nerve** which passes out of the pelvis above piriformis and supplies in addition the **tensor fasciae latae**. The latter muscle, which has a similar action to that of gluteus medius and minimus, arises from the anterior part of the iliac crest and inserts into the thick ilio–tibial tract of the fascia lata.

Stand at ease with your hip muscles as relaxed as possible and put your hands at your side. Now stand on one leg. You will feel the glutei and tensor fasciae latae contract firmly on the side of the leg on which you stand. You will also feel the pelvis rise slightly on the side on which you do not stand. The alternate contraction and relaxation of the gluteal muscles, and the rise and fall of the opposite side of the pelvis, can be felt on walking forward at a slow gait. When you stand on one leg, the pelvis pivots

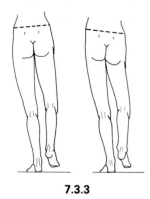

7.3.3

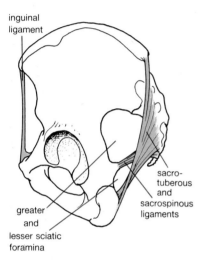

inguinal
ligament

sacro-
tuberous
and
sacrospinous
ligaments

greater
and
lesser sciatic
foramina

7.3.4

on the hip joint and the gluteal muscles contract to balance the body weight and elevate the pelvis, so that the non weight–bearing leg is enabled to swing in the normal stride. Now suppose the system is upset by paralysis of the gluteal muscles or by congenital dislocation of the hip, the normal balance of the hip joint will no longer work. When the patient stands on a leg in which the glutei are paralysed, or the hip congenitally dislocated, instead of the opposite side of the pelvis rising, it falls. The patient is said to have a positive Trendelenberg sign (**7.3.3**). Also, when the patient walks, the gait is characteristically waddling and is described as a Trendelenberg gait.

Before studying the other muscles of the gluteal region examine the bony pelvis with its attached ligaments (**7.3.4**). Note the **sacro–tuberous ligament** attached to the side of the sacrum and the ischial tuberosity and examine carefully the **sacro–spinous ligament** which lies deep to the former ligament and is attached to the side of the sacrum and the ischial spine. Two foramina or openings into the pelvis are thus formed, one above the sacrospinous ligament, the **greater sciatic fora-men**; the other below the same ligament, the **lesser sciatic foramen**.

Take note of the structures which pass through the greater sciatic foramen: the most prominent is **piriformis** (**7.3.5**) which, having arisen from the anterior aspect of the sacrum, passes through the greater sciatic foramen into the gluteal region. It

runs laterally, posterior to the hip joint, and inserts into the upper border of the greater trochanter under cover of gluteus medius. You should already have identified the **superior gluteal vessels** which are branches of the internal iliac vessels and the **superior gluteal nerve** which originates from the sacral plexus, both these structures emerge from the pelvis above piriformis. The large **sciatic nerve** (L4, 5, S1, 2, 3) which also originates from the sacral plexus, passes through the greater sciatic foramen below piriformis and enters the thigh lying in the lower, inner quadrant of the buttock and separated from the hip joint only by the small gluteal muscles. The **posterior cutaneous nerve of the thigh** (with a distribution as its name suggests) lies superficial to it while the **inferior gluteal nerve** and blood vessels which also originate from the sacral plexus and internal iliac vessels respectively, pass directly into the gluteus maximus (**7.3.6**).

In order to orientate yourself, examine a bony pelvis in which the sacrotuberous and sacrospinous ligaments are attached and then locate, on the prosected part, both the ischial spine and its attached sacrospinous ligament. Having done so, examine the **pudendal nerve** and **internal pudendal artery** and the **nerve to obturator internus** all of which arise in the pelvis either from the sacral plexus or internal iliac artery; they leave the pelvis by the greater sciatic foramen and cross the gluteal aspect of the sacrospinous ligament to enter the lesser sciatic foramen. The nerve to **obturator internus** supplies the muscle as it takes origin from the walls and fascia of the obturator foramen within the pelvis. On passing through the lesser sciatic foramen the pudendal nerve and vessels lie in the perineal region.

Examine the tendon of the obturator internus which passes out of the lesser sciatic foramen to run laterally deep to the sciatic nerve, thereby separating it from the hip joint. The tendon inserts into the greater trochanter and although it appears to be fleshy this is because it is flanked by the **superior and inferior gemelli** which arise from the bony borders of the lesser sciatic foramen and insert into the tendon of the obturator internus.

Identify **quadratus femoris** which runs horizontally from the lateral border of the ischial tuberosity to the intertrochanteric crest of the femur. Its nerve of supply emerges from the greater sciatic foramen deep to the sciatic nerve and enters the deep aspect of the muscle.

The buttock is often used for deep injections; they should be given into the upper outer quadrant of the buttock to avoid damaging the sciatic nerve (**7.3.7**).

While studying the gluteal region it is a convenient time to appreciate the danger of a posterior dislocation of the hip joint. If a passenger, not wearing a seat belt, is sitting in a car which is brought to a sudden halt, the passenger is projected forwards. With the hip flexed to a right angle, and the full force of the blow taken on the knees, the head of the femur may be pushed out backwards from the acetabulum, quite commonly taking with

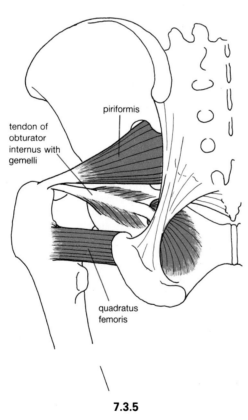

piriformis

tendon of
obturator
internus with
gemelli

quadratus
femoris

7.3.5

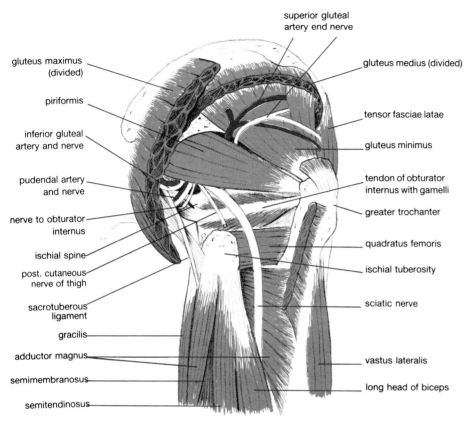

gluteus maximus (divided)
piriformis
inferior gluteal artery and nerve
pudendal artery and nerve
nerve to obturator internus
ischial spine
post. cutaneous nerve of thigh
sacrotuberous ligament
gracilis
adductor magnus
semimembranosus
semitendinosus

superior gluteal artery end nerve
gluteus medius (divided)
tensor fasciae latae
gluteus minimus
tendon of obturator internus with gamelli
greater trochanter
quadratus femoris
ischial tuberosity
sciatic nerve
vastus lateralis
long head of biceps

7.3.6

it a sharp fragment of the lip of the acetabulum. Since the sciatic nerve runs downwards and outwards from the greater sciatic foramen across the posterior aspect of the acetabulum, the nerve may very easily be contused or even lacerated.

Requirements:
    Articulated skeleton
    Pelvis with sacrotuberous and sacrospinous ligaments attached
    Prosections of gluteal region showing gluteus maximus in place, and gluteus maximus reflected to reveal the deep muscles of the gluteal region with the sciatic nerve and pudendal nerve.

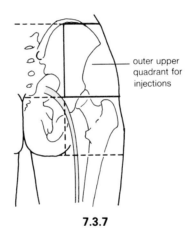

outer upper quadrant for injections

7.3.7

Seminar 4

Muscles and movements of the thigh

The **aim** of the following two seminars is to study the various groups of muscles which act on the hip, knee, and ankle joints.

Segmental innervation of muscles

Movements at a joint are usually controlled by four continuous segments of the spinal cord; the upper two segments innervating one movement and the lower two innervating the opposite movement.

Make all the following movements yourself, at the same time considering the spinal nerves which are involved.

Table 2 Segmental nerve supply to movements of the lower limb

Joint	Muscle action	Root supply	Muscle action	Root supply
Hip	Flexors adductors med. rotators	L1,2,3	Extensors abductors lat. rotators	L4,5, S1
Knee	Extensors	L 3,4	Flexors	L 5, S1
Ankle	Dorsi–flexors	L 4,5	Plantar–flexors	S1, 2
Subtalar	Invertors	L 4,5	Evertors	L 5, S1
Foot	Toe extensors	L 5, S1	Toe long flexors	S 2
			Small muscles of sole	S3

A. Prosections

a. Muscles of the front of thigh (7.4.2)

Examine the conjoined tendons of **iliacus** and **psoas major** (7.4.1) which pass deep to the inguinal ligament to be inserted into the lesser trochanter of the femur. Iliacus arises from the iliac fossa, psoas major from the bodies and transverse processes of the lumbar vertebrae. They both act as powerful flexors of the hip joint i.e. they flex the thigh on the body and, when the legs are weight–bearing, the body on the thigh.

Identify the four parts of the **quadriceps femoris** (7.4.3): the **rectus femoris** which originates from the anterior inferior iliac spine and the ischium immediately above the acetabulum; the **vastus intermedius** which arises from the front and sides of the femoral shaft; the **vastus medialis** which arises from the medial aspect of the inter-trochanteric line and the linea aspera; the **vastus lateralis** which arises from the lateral aspect of the intertrochanteric line, the base of the greater trochanter and the linea aspera. All four muscles' bellies are inserted into the upper margin of the patella, the tendon of which is attached to the tibial tubercle. Vastus medialis has, in addition, an attachment to the medial surface of the patella.

This muscle group is supplied by the femoral nerve and it extends the knee joint.

Qu. 4A *Which muscle of the quadriceps group acts on the hip joint and what movement is produced?*

Qu. 4B *Why is the muscle attached to the medial surface of the patella so important?*

Examine the **sartorius** (7.4.4) which arises from the anterior superior iliac spine and crosses the thigh obliquely to insert into the anterior aspect of the medial condyle of the tibia. Sartorius, (the name if derived from the latin word for tailor) is a prime mover when one is sitting cross–legged. Note the **tensor fasciae latae** which arises from the anterior part of the iliac crest and passes down the lateral aspect of the thigh to be inserted into the ilio–tibial tract. This musculo–tendinous unit aids the gluteus medius and gluteus minimus in their actions and, like them, is supplied by the superior gluteal nerve. Iliacus, quadriceps femoris, and sartorius are all supplied by the **femoral nerve** (L2, 3, 4); psoas major from the roots of the lumbar plexus.

The extensor mechanism of the knee

The quadriceps femoris, the patella, and the ligamentum patellae are together referred to as 'the extensor mechanism of the knee', thus emphasizing the unity and function of the separate structures. The stability of the knee depends on it. If quadriceps femoris is weakened, the knee will very easily give way, causing the body to fall. The extensor mechanism can be put under great strain, for instance the tibial tubercle can be avulsed from its bed or the lower pole of the patella pulled off. Sometimes the patella may fracture and the retinaculae tear. These injuries usually occur in young age groups.

In the elderly, however, the quadriceps tendon can give way, so that the patient cannot extend the knee fully against gravity.

Qu. 4C *Examine 7.4.5. The patient complained of pain in the thigh during a rugger match. What has happened?*

Wasting of the quadriceps is an early physical sign of disease of the knee joint. This is often detectable first in the vastus medialis which is prominent above the inner aspect of the knee in muscular subjects. Using your partner compare the size of the muscular prominence on both sides and see if you detect any difference. Also measure the circumference of the thighs at two different levels above the patella. Often the dominant thigh (as for kicking) will be measurably greater than the other.

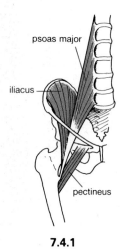

psoas major

iliacus

pectineus

7.4.1

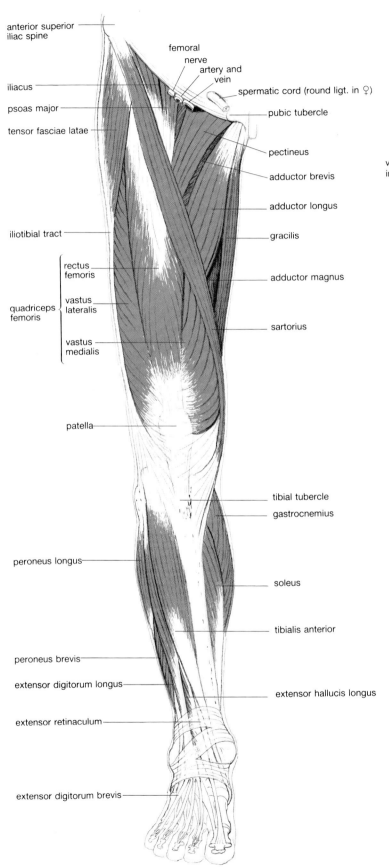

anterior superior
iliac spine

femoral
nerve

artery and
vein

iliacus

spermatic cord (round ligt. in ♀)

psoas major

pubic tubercle

tensor fasciae latae

pectineus

adductor brevis

adductor longus

iliotibial tract

gracilis

rectus
femoris

adductor magnus

quadriceps
femoris

vastus
lateralis

sartorius

vastus
medialis

patella

tibial tubercle

gastrocnemius

peroneus longus

soleus

tibialis anterior

peroneus brevis

extensor digitorum longus

extensor hallucis longus

extensor retinaculum

extensor digitorum brevis

7.4.2

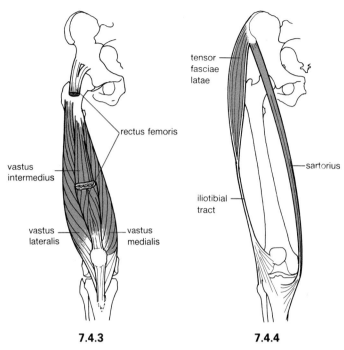

rectus femoris

vastus
intermedius

vastus
lateralis

vastus
medialis

7.4.3

tensor
fasciae
latae

sartorius

iliotibial
tract

7.4.4

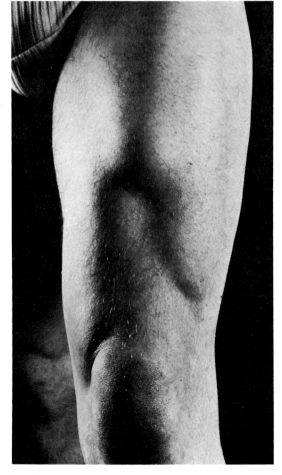

7.4.5

The line of pull of the quadriceps muscle is predominantly in line with the shaft of the femur. Since the femur is not vertical, this tends to pull the patella a little laterally, a force which is counteracted to some degree by the lower fibres of the vastus medialis arising from the lower end of the adductor magnus tendon and inserting into the upper medial edge of the patella. Also the lateral condyle of the femur projects more anteriorly than the medial condyle, acting as a guard against lateral displacement of the patella. Sometimes, however, if the patella is smaller and slightly higher than normal, or if there is a tendency to 'knock knee', the patella can tend to recurrent dislocation. It will move laterally to the extent that it jumps over the lateral condyle when the quadriceps is reflexly inhibited, and the patient's knee 'collapses.'

b. Muscles of the medial aspect of thigh (7.4.6, 7.4.7, 7.4.8)

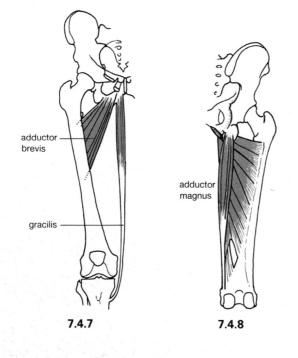

7.4.6   **7.4.7**   **7.4.8**

pectineus

adductor longus

adductor brevis

gracilis

adductor magnus

Examine **pectineus** which arises from the pectineal surface of the pubic bone and inserts into the femur below the lesser trochanter; the **adductor longus** and the **adductor brevis** which also arise from the pubic bone (the longus by a long tendon attached immediately below the pubic tubercle; brevis from the body and inferior ramus of the pubis) and pass medially and downwards to be attached to the linea aspera.

The **adductor magnus** has an adductor portion which arises from the ischio–pubic ramus and a hamstring part which arises from the ischial tuberosity. Its fibres pass medially and downwards to insert into the whole length of the linea aspera and the upper part of the medial supracondylar line; at this point it forms a fibrous arch through which the femoral neurovascular bundle passes from the anterior to the posterior compartment of the thigh. The distal tendinous end of the arch is attached to the **adductor tubercle**. The **gracilis** (an adductor) runs from the pubic arch to the medial aspect of the tibial shaft at its upper end.

This entire adductor muscle group is supplied by the **obturator nerve** (L2, 3, 4). The nerve, in passing out of the pelvis, travels through the obturator foramen and pierces the **obturator externus** which is attached to its outer bony margins. The tendon of the obturator externus passes laterally below and behind the neck of the femur to be attached to the greater trochanter and can therefore rotate the femur laterally.

Qu. 4D   *What actions other than adduction might any of these muscles perform?*

Qu. 4E   *If it was necessary to locate the pubic tubercle accurately (in order to determine whether a hernia was of the inguinal of femoral variety) how, with a knowledge of the muscles you have just identified, would you do so?*

c. Muscles of the back of the thigh (7.4.9, 7.4.10, 7.4.11)

The group of muscles which lie in the posterior compartment of the thigh are known as '**the hamstrings**'. They arise from the ischial tuberosity and pass down the back of the leg bound tightly together by a sleeve of tough fascia (fascia lata) which surrounds the entire thigh. The **semimembranosus** originates by a membranous aponeurosis and is inserted into the posterior aspect of the medial condyle of the tibia; the **semitendinosus**, as its name suggests, ends in a long tendon which passes downwards over the former muscle and around the medial side of the knee to be inserted into the upper end of the medial aspect of the tibial shaft. The **long head of biceps femoris** joins its **short head** which arises from the linea aspera and the lateral supracondylar ridge of the femur; the combined muscle ends in a tendon which passes downwards to insert into the head of the fibula on either side of the lateral ligament of the knee joint. That part of the adductor magnus which arises from the ischial tuberosity is also considered as a hamstring. This entire group of muscles extends the hip joint but flexes the knee joint; the group is supplied by the **sciatic nerve** (L4, 5, S1, 2, 3).

Qu. 4F   *Which nerves supply adductor magnus?*

Elicit a knee jerk: get your partner to sit down and relax, with one leg crossed over the other; then, using the patellar hammer, tap the quadriceps tendon of the crossed leg. Record your results and explain the physiological mechanisms underlying the reflex response.

Requirements:
     Articulated skeleton and separate bones of lower limb
     Prosections of anterior compartment of the thigh, medial compartment of the thigh, and posterior compartment of the thigh.

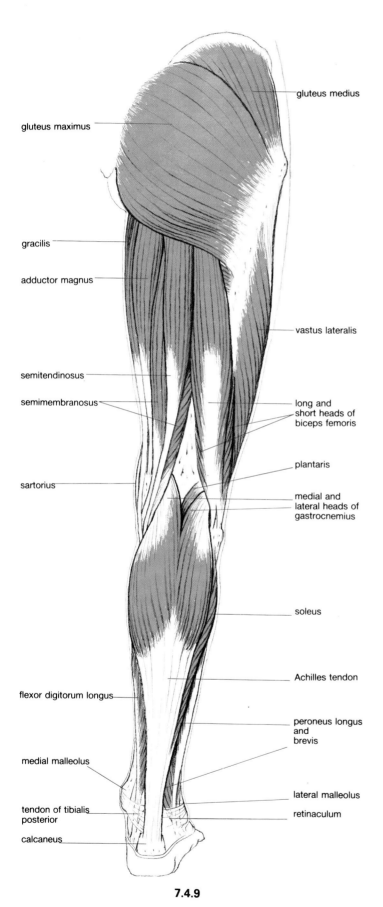

gluteus medius

gluteus maximus

gracilis

adductor magnus

vastus lateralis

semitendinosus

semimembranosus

long and
short heads of
biceps femoris

plantaris

medial and
lateral heads of
gastrocnemius

sartorius

soleus

Achilles tendon

flexor digitorum longus

peroneus longus
and
brevis

medial malleolus

lateral malleolus

tendon of tibialis
posterior

retinaculum

calcaneus

7.4.9

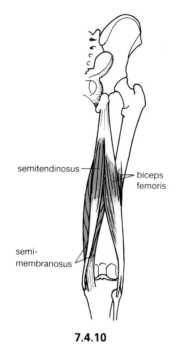

semitendinosus

biceps
femoris

semi-
membranosus

7.4.10

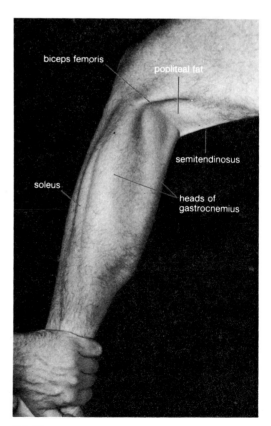

biceps femoris

popliteal fat

semitendinosus

soleus

heads of
gastrocnemius

7.4.11

Seminar 5

Muscles and movements of the leg

A. Prosections

a. Muscles of the front of leg and dorsum of foot

Tibialis anterior, extensor digitorum, and extensor hallucis longus (**7.4.2**) all arise from the anterior aspect of the shaft of either the tibia or the fibula and the interosseous membrane. Extend (dorsi-flex) your own foot and note that the most medial tendon is that of **tibialis anterior** (**7.5.1**). Turn to the cadaver and trace the tendon from its muscle belly to its attachment on the medial aspect of the medial cuneiform bone and the adjoining part of the base of the first metatarsal bone. Now extend your big toe (hallux) and note the tendon of **extensor hallucis longus** (**7.5.3**); identify it on the cadaver (**7.5.1**) and look for its insertion into the base of the distal phalanx of the big toe. The tendons of **extensor digitorum longus** (**7.5.2**) are the most laterally placed on the dorsum of the foot and broaden into dorsal expansions on the lateral four toes; the central part of each tendon attaches to the base of the 2nd phalanx while the lateral parts insert into the base of the distal phalanx. An extension of this muscle, known as peroneus tertius, is inserted into the dorsal aspect of the 5th metatarsal. The **extensor digitorum brevis** arises from the dorsal surface of the calcaneum and fills out what otherwise would be a hollow on the lateral aspect of the dorsum of the foot. Three of its tendons join the dorsal extensor expansion while the one to the big toe (sometimes termed extensor hallucis brevis) has a separate insertion into the base of the proximal phalanx. This muscle group is supplied by the **deep peroneal nerve**.

Note the **extensor retinaculum** which keeps the tendons in place on the front of the ankle (**7.6.6**). The fascia which encloses the anterior and posterior compartments is very firm and swelling of its contents can cause severe pressure effects (see Seminar 7).

Qu. 5A *What are the actions of all these muscles and in particular those of tibialis anterior?*

b. Muscles of the lateral aspect of leg

The lateral compartment of the leg contains **peroneus longus** (**7.5.4**) which arises from the upper two thirds of the lateral surface of the fibula and **peroneus brevis** which lies beneath the longus, and arises from the lower two thirds of the same bony surface. Peroneus longus has a long tendon which runs behind the lateral malleolus. After grooving the cuboid bone the tendon passes across the sole of the foot to insert close to tibialis

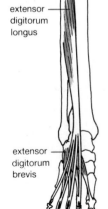

tibialis anterior

extensor hallucis longus

7.5.1

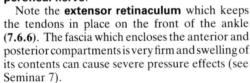

extensor digitorum longus

extensor digitorum brevis

7.5.2

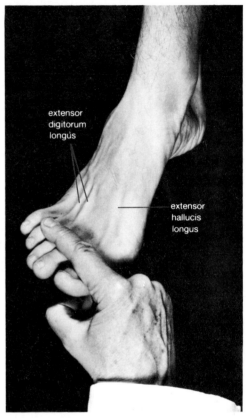

extensor digitorum longus

extensor hallucis longus

7.5.3

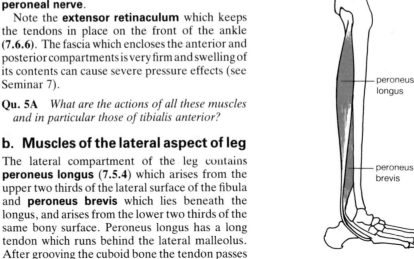

peroneus longus

peroneus brevis

7.5.4

anterior on the medial cuneiform bone and the base of the 1st metatarsal, thus providing a stirrup–like support for the foot.

Qu. 5B *What other actions do these muscles have?*

The tendon of peroneus brevis also passes behind the lateral malleolus (**7.5.5**) and is inserted into the base of the 5th metatarsal bone. Note the **peroneal retinaculum** which holds these tendons in place behind the lateral malleolus. Peroneus longus and brevis are supplied by the **superficial peroneal nerve**.

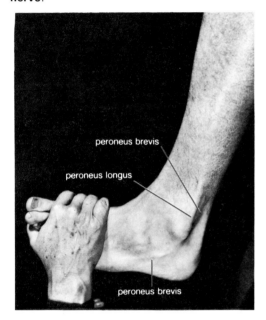

7.5.5

c. Muscles of the back of the leg

Superficial calf muscles (7.4,9)
The **gastrocnemius** (**7.5.6**) arises by two fleshy bellies, one from each of the condyles of the femur; these bellies unite to be inserted by a large tendon, the **tendo calcaneus** or **Achilles tendon**, into the posterior surface of the calcaneus. The **popliteus** (**7.5.7**) is triangular in shape and arises from the upper part of the posterior surface of the tibia. Its tendon, which forms the apex of the triangle, inserts into a pit on the outer aspect of the lateral femoral condyle, within the capsule of the joint. The tendon separates the lateral meniscus of the knee joint, to which it has an attachment, from the lateral ligament. On contraction it is currently considered to rotate the femur laterally on the tibia thereby 'unlocking' the fully extended knee. **Soleus** arises from a horseshoe–shaped origin which spans both the tibia and fibula below the insertion of popliteus, its tendon blending with the tendo calcaneus. Perforating veins pass through this muscle and it is therefore an important component of the venous pump.

Look for and feel, on your partner's leg, the separate bellies of gastrocnemius and soleus.

Qu. 5C *What are the actions of the superficial calf muscles?*

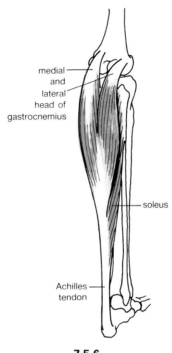

7.5.6

The tendo calcaneus transmits very considerable forces, and is the thickest tendon in the body. In middle age, if a person suddenly takes part in relatively unaccustomed exercise, the tendon may snap, usually at the site of some degenerative changes. The typical story is of a middle–aged man or woman starting to play badminton again after a long lay off. He suddenly feels as if his partner has hit him over the heel, and he may even hear the snap, 'like a pistol shot!'

Qu. 5D *What effect would such a injury have on the function of the foot*
   a) when free of the ground?
   b) when weight bearing?

Ask your partner to kneel on a chair and to relax his calf muscle. Squeeze the calf from side to side. With an intact tendo calcaneus the foot moves

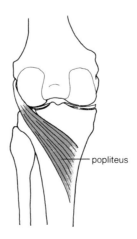

7.5.7

passively into plantar flexion at each squeeze and returns to the normal position when the pressure is relaxed. If the tendon is ruptured no such movement will occur. This is the calf 'squeeze test' for a rupture of the tendon. The usual treatment is to resuture the ruptured tendon as soon as possible to avoid its healing in a lengthened position, which would weaken the 'push–off' of the foot in walking.

Deep muscles of the calf (7.5.10)
Lying deeply under cover of the superficial muscles are the **flexor hallucis longus**, (7.5.8) **flexor digitorum longus** (7.5.8) and **tibialis posterior** (7.5.9), which take origin from the posterior aspect of the fibula, the tibia, and the interosseous membrane and surrounding bone respectively. They are inserted respectively into the distal phalanx of the hallux, the distal phalanges of the lateral four toes, and the navicular.

At the ankle joint they are held in position by the **flexor retinaculum** (see **7.6.6**) which is attached to the medial malleolus, to the sustentaculum tali, and to the medial surface of the calcaneum below the sustentaculum.

All muscles of the flexor compartment of the calf are supplied by the **tibial nerve**.

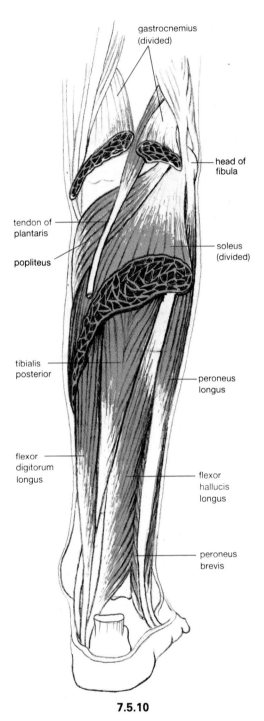

7.5.10

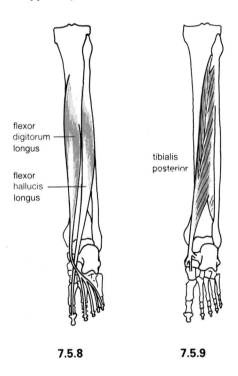

7.5.8   7.5.9

Requirements:
   Articulated skeleton and separate bones of the lower limb
   Prosections of the anterior compartment of the leg and dorsum of the foot; of the lateral compartment of the leg; of the superficial muscles of the calf; and of the deep muscles of the calf.

Seminar 6

The foot

The **aim** of this seminar is to study the function of the foot. For this purpose the general architecture of the foot should be understood and appreciated but it is not necessary to learn details of the origins and insertions of all its small muscles.

A. Prosections

Muscles of the sole of the foot

The fat of the sole and heel is thick and granular due to fine fibrous septa which pass through it from the skin to the plantar aponeurosis.

Qu. 6A *What is the functional importance of this arrangement?*

The plantar aponeurosis lying deep to the subcutaneous tissue is exceptionally thick and strong and consists of both longitudinal and transverse bands of fibres; it is attached posteriorly to the calcaneus while anteriorly it divides into five slips which extend forward into the toes to become attached to the sides of the proximal phalanx.

Examine the muscle and tendon of the four layers of the sole of the foot:

Layer 1. (**7.6.1**) Beneath the aponeurosis the **abductor hallucis** arises from the medial tubercle of the calcaneus and the surrounding fascia and inserts into the medial side of the base of the proximal phalanx of the big toe. It abducts and flexes the big toe. The **abductor digiti minimi** arises from both tubercles of the calcaneum and inserts on the lateral aspect of the base of the proximal phalanx of the little toe.

The **flexor digitorum brevis** arises from the medial calcaneal tubercle and the aponeurosis and divides into four tendons which split to allow the tendons of flexor digitorum longus to pass through before being inserted into the sides of the middle phalanx. The tendons, as in the hand, pass through a protective fibrous arch into the toes.

Qu. 6B *Which muscle tendons in the hand behave in a similar fashion to those of flexor digitorum brevis?*

Layer 2. (**7.6.2**) Beneath the muscles of layer 1 lie the tendons of two of the deep calf muscles. The **flexor digitorum longus tendon** divides into four slips which insert into the base of the terminal phalanx of the four lateral toes; the **flexor hallucis longus tendon**, surrounded by a fibrous sheath, grooves the back of the talus and the sustentaculum tali, runs medially along the sole and inserts into the base of the terminal phalanx of the big toe. Also lying in this so–called 2nd layer of the sole of the foot is **flexor accessorius** which arises largely from the medial concave surface of the calcaneus

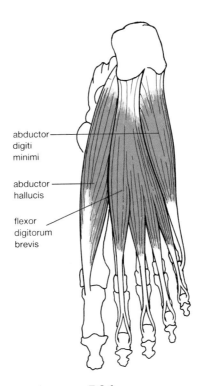

abductor digiti minimi

abductor hallucis

flexor digitorum brevis

7.6.1

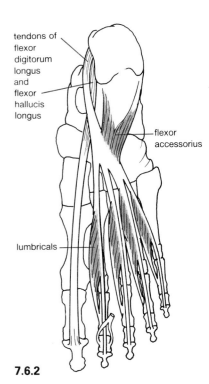

tendons of flexor digitorum longus and flexor hallucis longus

flexor accessorius

lumbricals

7.6.2

and is inserted into the deep surfaces of the flexor digitorum longus tendons. The action of flexor accessorius is to convert the oblique pull of the longus tendons into a direct antero–posterior one. The four **lumbrical muscles** arise from the tendons of flexor digitorum longus; as in the hand, they cross the metatarso–phalangeal joints on the tibial side to insert into the dorsal extensor expansion and thus flex the metatarso-phalangeal joints.

Qu. 6C *Grip the floor with your toes; how important is this action?*

Layer 3. (**7.6.3**) Beneath the 2nd muscular layer of the sole lie the following three muscles: **flexor hallucis brevis**, which arises from the ligament on the medial side of the sole, has two bellies, the tendons of which insert into either side of the base of the proximal phalanx of the big toe; sesamoid bones can be found within these tendinous insertions. The **adductor hallucis** has both a transverse and an oblique head, the common tendon of which is inserted into the lateral side of the base of the proximal phalanx and lateral sesamoid bone of the big toe; **flexor digiti minimi brevis** arises from the plantar surface of the base of the 5th metatarsal bone and is inserted into the lateral side of the base

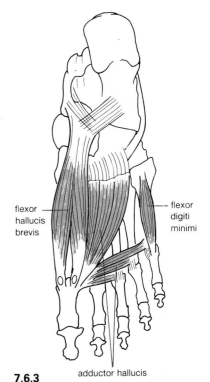

flexor hallucis brevis

flexor digiti minimi

adductor hallucis

7.6.3

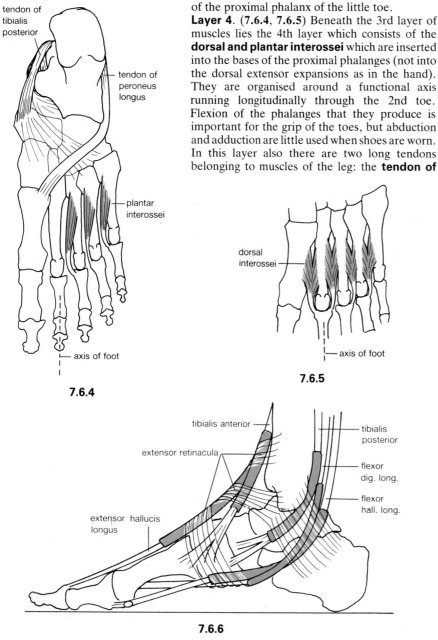

7.6.4

dorsal interossei

axis of foot

7.6.5

of the proximal phalanx of the little toe.

Layer 4. (7.6.4, 7.6.5) Beneath the 3rd layer of muscles lies the 4th layer which consists of the **dorsal and plantar interossei** which are inserted into the bases of the proximal phalanges (not into the dorsal extensor expansions as in the hand). They are organised around a functional axis running longitudinally through the 2nd toe. Flexion of the phalanges that they produce is important for the grip of the toes, but abduction and adduction are little used when shoes are worn. In this layer also there are two long tendons belonging to muscles of the leg: the **tendon of**

tibialis posterior which passes behind the medial malleolus to insert into the tuberosity of the navicular bone, and the **tendon of peroneus longus** which inserts into the medial cuneiform and the base of the 1st metatarsal. Having identified all these muscles and tendons of the foot, many of which are reminiscent of those found in the hand, it will be much easier for you to consider the arch of the foot and its functional importance.

The **medial plantar branch of the tibial nerve** supplies abductor hallucis; flexor digitorum brevis, flexor hallucis brevis, and the first lumbrical. The **lateral plantar nerve** supplies the remaining muscles.

Qu. 6D   *Compare this pattern of innervation with that in the hand. Which nerve is comparable to the median nerve?*

The synovial tendon sheaths surrounding tendons at the ankle are illustrated in **7.6.6, 7.6.7**. Compare this arrangement with that at the wrist (see **6.12.4, 6.12.5**).

The arch of the foot (7.6.8)

The arch of the foot enables the weight of the body to be equally distributed between the medial and lateral tubercles of the calcaneus posteriorly and the heads of the lateral four metatarsals and two sesamoid bones of the first metatarsal anteriorly. The basic structure of the arch, which consists of a number of small bones united by many articular joints, allows for considerable flexibility and spring, and this in turn facilitates walking and running. In addition, the toes help to provide leverage for propelling the body forwards and by their gripping action they also help to provide stability and balance both in standing and during movement.

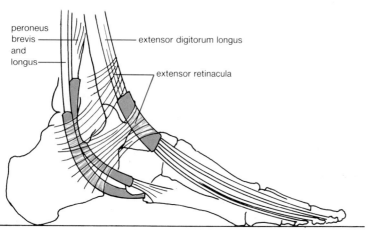

tibialis anterior

extensor retinacula

extensor hallucis longus

tibialis posterior

flexor dig. long.

flexor hall. long.

7.6.6

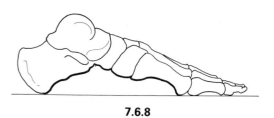

7.6.8

peroneus brevis and longus

extensor digitorum longus

extensor retinacula

7.6.7

Although in most anatomical texts you will find that the arch of the foot is analysed in terms of medial, lateral, and transverse components, it is clear that the basic structure consists of individual bones of the foot which are united by the relevant joint capsules and reinforced by **dorsal** and **plantar ligaments**. There are, however, other **features which strengthen and maintain the arch**. Perhaps the most important of these, and certainly during movement, are the **small muscles of the sole of the foot**. In considering further supports of the arch the **long tendons of the calf muscles (7.6.9)** are of importance, in particular the tibialis anterior tendon which is inserted alongside the peroneus longus tendon to form a strong tendinous stirrup. In addition the **long and short plantar ligaments (7.2.30)** extend from the lateral tubercle

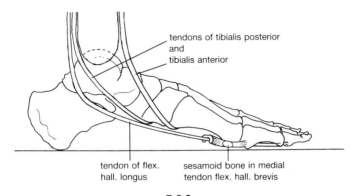

tendons of tibialis posterior
and
tibialis anterior

tendon of flex.
hall. longus

sesamoid bone in medial
tendon flex. hall. brevis

7.6.9

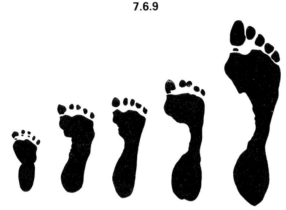

7.6.10 Development of footprints from six-months to adulthood.

of the calcaneus to the distal carpal bones to strengthen the lateral aspect of the arch and, in doing so, form a fibrous tunnel for the peroneus longus tendon. On the medial side of the sole there is a potential weakness in that weight transmitted through the tibia and fibula will tend to thrust the head of the talus between the calcaneus and the navicular bones. The presence of a strong **spring ligament (7.2.30)**, which extends between the sustentaculum tali and the navicular bone, supports the head of the talus and prevents this possibility. When standing passively without movements of the foot, the ligaments are the most important supporting features of the bony arch.

B. Living anatomy

A baby's foot does not appear to possess the medial longitudinal arch. When the child begins to walk, towards the end of the first year or beginning of the second, the relatively large amounts of fat regress and the arch begins to appear. However, the normally 'flat' appearance may persist through later life. Ask your partner to remove his shoes and socks, and to stand relaxed with his feet slightly apart. Look at his feet from in front and from behind. You will see that the medial arch is rather flattened, but off the surface of the floor (**7.6.10**). If you compare several pairs of feet you will probably note some variations in the height of the arch.

You will also note that the heel is somewhat everted, sloping downwards to the inner side of the foot. Now ask your partner to stand on his toes. The heel moves from slight eversion to slight inversion, and the medial arch becomes much more pronounced.

Get your partner to take a step forward, and notice, as the heel of the weight–bearing foot rises from the ground, how the big toe is necessarily passively dorsiflexed at the metatarsophalangeal joint. Look at the skeleton of a foot and notice how the proximal end of the second metatarsal is partially gripped between the medial and lateral cuneiform bones. It sometimes happens after a long march on a soft, untrained recruit's foot, or even in a seasoned long distance walker's, that a fracture can occur through the distal end of the shaft of the second metatarsal. There is sudden pain and a limp, but at first an x–ray does not show any fracture. Within a week or two, however, new bone can be felt, and seen on x–ray, at the fracture site. Bone will fracture if it is exposed to a single severe stress, or alternatively if it is exposed to continually repeated more minor stress, as in the metal of an aircraft's wing. Such a fracture is known as fatigue fracture, and the 'march fracture' is one such. It is thought that the second metatarsal has more stress applied to its distal shaft, because of the lesser mobility of its base.

Requirements:
 Articulated skeleton and separate bones of the
  foot
 Prosections of the four layers of the sole of the
  foot; and of the ligaments of the foot.

Seminar 7

Blood supply and lymphatics of the lower limb

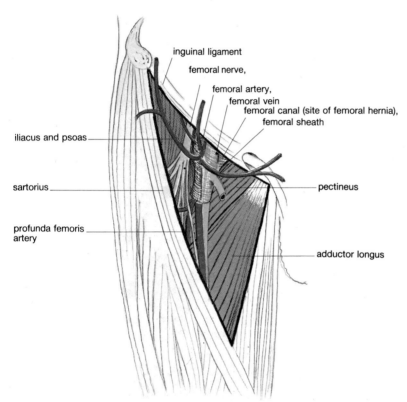

7.7.1

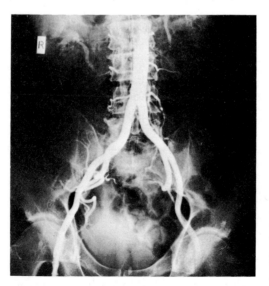

7.7.2

The **aim** of this seminar is to study the blood supply and lymphatic drainage of the lower limb with the aid of your partner and prosected parts.

A. Living anatomy (7.7.3)

It is important that you should be able to feel the pulses of the normal living leg since their absence may signify arterial disease. Ask your partner to lie on a couch and feel first for the pulsation of the **femoral artery** in the groin. This is usually easy to find. Press gently with one, or at most two, finger tips just below the mid–point of the inguinal ligament (halfway between the pubic tubercle and the anterior superior iliac spine).

Feel for the **dorsalis pedis artery** in the foot which lies at the mid point between the medial and lateral malleoli. It continues on the dorsum of the foot lateral to the tendon of the extensor hallucis longus until it reaches the space between the proximal ends of the first and second metatarsals; here the artery may also be felt before it penetrates to the sole.

Feel for the **posterior tibial artery** behind the medial malleolus, where it lies under the flexor retinaculum.

Ask your partner to flex his knee to a right angle and insinuate your finger tips into the popliteal fossa behind the knee, pressing gently towards the back of the knee joint. You may be able to feel the pulsation of the **popliteal artery**, though this is sometimes difficult to find. It may be easier to feel the artery with your partner lying prone. Do not forget that the leg also receives blood supply via the gluteal arteries in the buttock, but these are deep and their pulsations cannot be felt.

B. Prosections

Arterial supply of the lower limb

When the **external iliac artery (7.7.2, 7.7.3)** passes under the mid–point of the inguinal ligament to enter the thigh, it is called the **femoral artery (7.7.1)**. As it enters the thigh it lies in a region known as the **femoral triangle**. Examine the boundaries of this triangle which are formed by the inguinal ligament above and the medial borders of the sartorius and adductor longus muscles below; thus the adductor longus (medially) and the pectineus and iliacus and psoas (laterally) form its floor. The femoral artery is accompanied on its

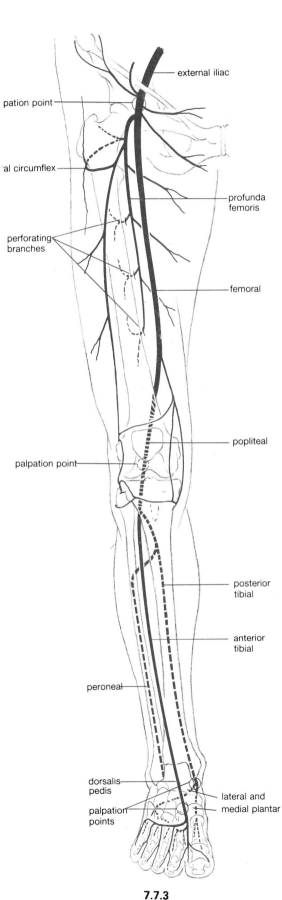

external iliac

pation point

al circumflex

profunda
femoris

perforating
branches

femoral

popliteal

palpation point

posterior
tibial

anterior
tibial

peroneal

dorsalis
pedis

palpation
points

lateral and
medial plantar

7.7.3

7.7.4

medial side by the **femoral vein** and both lie within separate fascial compartments of a funnel–like prolongation of the abdominal fasciae. This protrusion of the abdominal fascial envelope beneath the inguinal ligament is known as the **femoral sheath**. As the femoral artery lies in the femoral triangle it gives off a large branch, the **profunda femoris** artery, which passes deeply between adductor longus and adductor magnus. This vessel gives off a number of branches: a medial and a lateral **circumflex femoral artery,** which encompass the femur to supply the adjacent muscles, and **perforating arteries** which pass through the adductor muscles to supply the muscles of the medial and posterior compartments and the skin of the thigh.

Follow the femoral artery as it leaves the femoral triangle and, passing deep to the sartorius, enters an intermuscular canal (the subsartorial or adductor canal) on the medial aspect of the thigh. At the inferior end of this canal the artery passes through an arch in the muscle fibres of adductor magnus to reach the back of the knee joint—an area which is usually termed the 'popliteal fossa'. The artery, which you have felt, probably with difficulty in this position, is then renamed the **popliteal artery** (**7.7.5**) and gives off branches of supply to the knee joint. As in other areas where movements take place, anastomoses are common and can be found between smaller arteries which encompass the lower end of the femur and the upper end of the tibia (**7.7.5**). After it leaves the popliteal fossa, the popliteal artery divides into **anterior and posterior tibial arteries**.

Identify and follow the **anterior tibial artery** as it passes between the tibia and fibula above the interosseous membrane and descends on its anterior surface to supply the muscles of the anterior compartment of the leg. As it reaches the ankle it gives off **medial and lateral malleolar branches**, and when it crosses the front of the ankle joint it is known as the **dorsalis pedis artery** (**7.7.6**). Trace this artery as it passes forwards over the dorsum of the foot to reach the interval between the 1st and 2nd metatarsal bones through which it passes to reach the sole of the foot. Just before it passes into

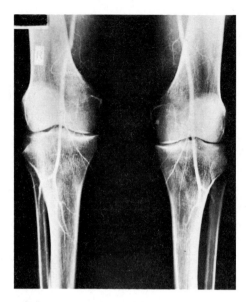

7.7.5

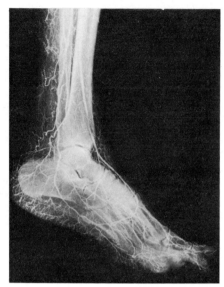

7.7.6

the sole, a **transverse metatarsal** branch is given off which runs laterally and supplies small paired branches to the toes.

Follow the **posterior tibial artery** which supplies the superficial and deep muscles of the calf and passes with the long tendons of this compartment around the back of the medial malleolus. Under cover of the flexor retinaculum it divides into the **medial and lateral plantar arteries** which enter the sole of the foot. The medial plantar artery is small and runs along the medial side of the foot. The lateral plantar artery runs obliquely across the sole of the foot between the 1st and 2nd layers of muscles towards the base of the 5th metatarsal. Here, the artery dips deeply and runs horizontally across the interossei, giving off **metatarsal branches** and ends by anastomosing with the dorsalis pedis artery (which has reached the sole from the dorsum of the foot) thereby forming a **plantar arterial arch**. A **peroneal artery** is also given off from the posterior tibial artery. It runs downwards with flexor hallucis longus and may take over the distribution of the dorsalis pedis artery.

Qu. 7A   *Are similar arterial features to those described above found in the palm of the hand?*

Qu. 7B   *If a patient who complained of intermittent pain in the legs (intermittent claudication) during and after exercise also has a toe which was becoming gangrenous, what would you deduce? How might such a patient be treated surgically?*

Look again at the prosection of the lower leg and examine the anterior compartment of the leg. Notice that the muscles, tibialis anterior, extensor hallucis longus, extensor digitorum longus, and peroneus tertius are enclosed in a compartment with firm inextensible walls. These walls consist of the deep fascia of the leg, the tibia, the interosseous membrane, the fibula, and the fascia separating the compartment from the peroneus longus and brevis. This compartmentation of muscle in the leg is an advantage to the peripheral venous circulation

in that a muscular pump is formed which forces blood, centripetally, against gravity, But if someone is suddenly subjected to unaccustomed severe and prolonged exercise, such as an army recruit, the muscles of the anterior compartment may swell from fatigue. The swelling may be such that, in the enclosed space, the venous return is obstructed; this leads to further swelling, and eventually to capillary stasis, and gangrene. The condition is termed the 'anterior compartment syndrome' of the leg and if it is not relieved it can be crippling. If the process is diagnosed early enough, relief can be obtained by dividing the deep fascia along the whole length of the compartment.

Qu. 7C   *Could the same problem develop with the muscles of the posterior part of the lower leg?*

The blood vessels of the lower limb are innervated by sympathetic autonomic fibres. Preganglionic fibres originate from cell bodies situated in the grey matter of lower thoracic and highest lumbar segments of the spinal cord. Postganglionic fibres arise from the lumbar sympathetic chain and form a plexus around the external iliac artery. Other fibres from the lumbar and pelvic parts of the chain join the femoral, obturator, and sciatic nerves in order to reach the vessels and their branches.

The **superior** and **inferior gluteal arteries**, both branches of the internal iliac artery, supply the gluteal region and send small branches down into the posterior compartment of the thigh where they make anastomoses with branches of the femoral supply system.

C. Radiographs

Study the arteriograms (**7.7.5**, **7.7.6**) and label the various branches shown. After looking at the normal condition look at the arteriogram in **7.7.7**. The difference in appearance is due to the presence of atheroma, a cholesterol–containing deposit, in the tunica intima of the arterial wall. This narrows the lumen of the vessel, sometimes blocking it off completely. Gangrene of the distal limb might occur in such cases.

Study the arteriogram (**7.7.8**). What has happened? The leg suffered a gun shot injury, shattering the bone and destroying the popliteal artery. The fractured bones were stabilised as well as was possible, but gangrene of the foot would have occurred unless the blood supply had been restored surgically.

Anastomoses between arteries supplying the lower limb

Finally, consider the anastomoses that occur between arteries of the lower limb. An important anastomosis is found in the trochanteric fossa, from which vessels pass to the head of the femur. Here branches of the gluteal arteries anastomose with branches of the medial and lateral femoral circumflex arteries. A little lower down, the

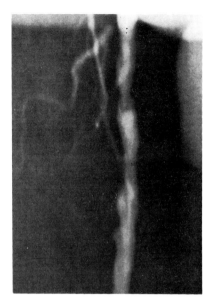

7.7.7

circumflex arteries also anastomose with gluteal arteries and the upper perforating branch of the profunda to form the 'cruciate' anastomosis. Still lower, the perforating arteries anastomose among themselves and with the genicular vessels. Thus a chain of anastomoses is created. You should consider ways in which the leg can derive arterial supply if the femoral trunk is blocked at different sites.

Venous drainage of the lower limb (7.7.9)

As with the upper limb there is a system of superficial veins which drain the skin and superficial fascia, and a system of deep veins, pairs of which accompany smaller arteries and are known as venae comitantes. Connecting these two sets, which are separated by the tough fascia lata, are a set of communicating veins.

Superficial drainage

On the dorsum of the foot there is a **dorsal venous arch** into which drain tributaries from the toes. The lateral aspect of the arch is drained by the **short saphenous vein** which runs behind the lateral malleolus and up the back of the calf to pierce the fascia of the popliteal fossa and drain into the popliteal vein (**7.7.10**). The medial side of the dorsal venous arch is drained by the **long saphenous vein** which passes in front of the medial malleolus where it can usually be seen. This long vein then ascends unsupported in the subcutaneous tissue along the medial aspect of the leg and thigh. At the top of the thigh, about 5cm below the pubic tubercle, it enters an opening in the fascia lata to join the **femoral vein** as it lies in the femoral triangle. At this point several smaller veins draining the iliac and peripheral regions join the femoral vein, and it is here that the saphenous vein can be conveniently tied off if it becomes varicose

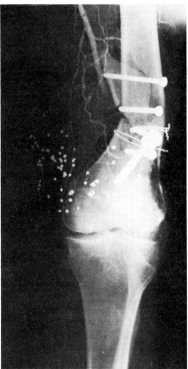

7.7.8

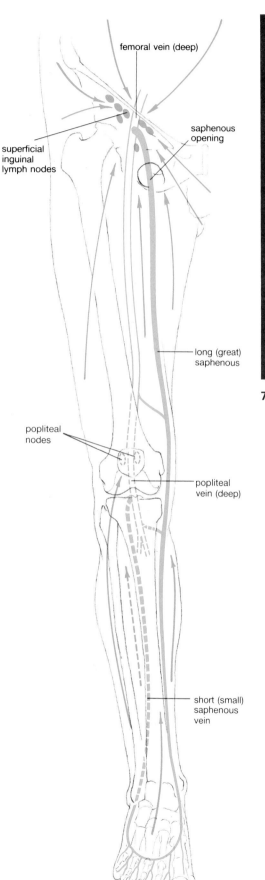

7.7.9

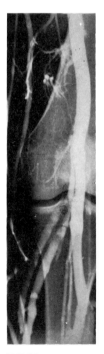

7.7.10

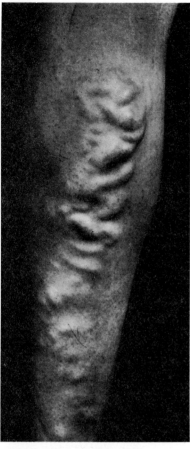

7.7.11

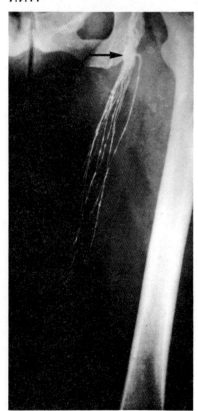

7.7.12

(**7.7.11**). Cut along the length of the saphenous vein and examine if for the presence of valves which divide up the column of blood and thus reduce the hydrostatic pressure. The long saphenous vein at the ankle is a favourite site for intravenous infusion when the arm veins are difficult to find.

Deep drainage

Two venae comitantes accompany each of the smaller arteries in the leg and eventually join to form the **popliteal vein**. This vessel along with the artery of the same name lies in the popliteal fossa. The vein passes through the intermuscular arch formed by the fibres of adductor magnus to reach the front of the thigh where it is renamed the femoral vein. The femoral vein while running upwards through the adductor canal and the femoral triangle receives many large tributaries including the profunda femoris vein. Having passed deep to the inguinal ligament into the pelvis, it is known as the external iliac vein. When you stand (and under various other circumstance), the venous pressure in the lower limb increases, as will the diameter of the femoral vein. This could lead to compression of the vein against the sharp pectineal part (so named because it is attached to the pectineal line) of the inguinal ligament. There is, however, potential space to accommodate the venous expansion on the medial side of the femoral sheath in the form of the femoral canal. The femoral canal is a continuation of the abdominal cavity (not the peritoneal cavity) and it is not surprising that occasionally, when the intra–abdominal pressure is increased, abdominal contents follow the path of least resistance and a femoral hernia results. Since the sharp pectineal part of the inguinal ligament lies on the medial side of the femoral canal, an increase in pressure within the hernia may lead to the blood supply (most usually the venous return) of its contents being obstructed. Such an event will lead to 'strangulation' of the hernia.

Qu. 7D *How, with your knowledge of anatomy, could you distinguish between an inguinal and a femoral hernia?*

The soleus muscle contains many veins within its mass which are difficult to demonstrate in the cadaver. The blood in these veins is especially liable to clot in circumstances of injury or disease in which immobilization and stasis of flow occurs. The presence of such a deep vein thrombosis can be demonstrated by means of a venogram. Radio–opaque dye is injected into a superficial vein of the

foot and a radiograph is taken shortly afterwards. **7.7.10** shows a normal venogram.

Communicating veins. A system of communicating veins joins the superficial veins, especially the great saphenous vein, to the deep veins running with the major arteries. These veins necessarily pierce the fascia lata and, if dilated, the point at which they pierce can be palpated as a tender defect in that layer. Major communicating veins are to be found just above and just below the knee joint, on its medial aspect. Search for these veins on a prosected specimen. The communicating veins have valves which normally permit blood to flow from the superficial to the deep veins. The deep veins are actively emptied of blood by the muscle pump and this arrangement allows blood to be indirectly pumped out of the superficial veins also.

Varicose veins are common (**7.7.11**). If, due to congenital or other reasons the venous valves become incompetent, the flow of blood in the communicating veins can be reversed. This can lead to stagnation in the superficial venous circulation, dilated superficial veins, and skin changes which can progress to ulceration.

Lymphatic drainage

(**7.7.9, 7.7.12**)

Again, as in the upper limb, the lymphatic drainage follows the general pattern of the superficial and deep venous drainage though there is very little communication between superficial and deep lymphatics. Superficial lymph glands (arrow; **7.7.12**) are found around the inguinal ligament and the entry of the long saphenous vein into the femoral vein. This superficial inguinal group of glands drains the skin and superficial fascia of the lower limb and the trunk below the waist linc. Since the lower part of the anal canal and the lower part of the external genitalia are developed from ectoderm, these areas also drain into the superficial inguinal lymph glands.

Requirements:
    Articulated skeleton of lower limb
    Prosections of the arterial supply to the lower limb showing femoral triangle and passage of femoral artery through adductor magnus to popliteal fossa; and posterior tibial arteries, medial and lateral plantar arteries; dorsalis pedis artery; gluteal arteries
    Arteriograms and venograms of the lower limb.

Seminar 8

Innervation of the lower limb: femoral and obturator nerves

The **aim** of these seminars is to trace, by dissection, the course and distribution of the femoral and obturator nerves and their branches. These are derived from the lumbosacral plexus which arises from lumbar roots deep to psoas and sacral roots on the posterior wall of the pelvis (**7.8.1**).

A. Living anatomy

With a skin pencil mark out the dermatome lines on your partner (**7.8.2**). Remember that each dermatome is supplied by a single spinal nerve. These dermatomes, however, overlap considerably except at the axial lines. When compared with the upper limb, the dermatome pattern of the lower limb is considerably distorted, largely because the limb has become medially rotated during development. As in the upper limb the central spinal nerve root of the plexus lies most peripherally i.e. the sole of the foot is supplied mainly by L5.

Qu. 8A *Which spinal nerve supplies the area of skin on which we sit?*

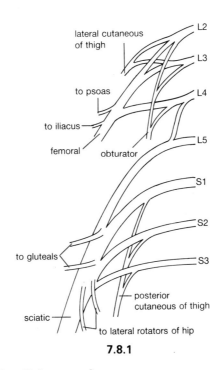

7.8.1

B. Dissection

a. The *femoral nerve* (7.8.1, 7.8.3, 7.8.4)

If you have not removed the skin from the lower limb before reaching this seminar, do so now. Locate the inguinal ligament, sartorius, and adductor longus and palpate the femoral artery and vein which lie in the femoral triangle.

Remove the thick deep fascia lata which ensheaths the thigh and forms a covering for the femoral triangle and identify the muscles of the anterior compartment. Locate the femoral sheath (the prolongation of abdominal fascia which forms a funnel around the femoral vessels) and push the tip of your little finger upwards alongside the medial aspect of the femoral vein. Your finger should now be within the femoral canal. Palpate the sharp pectineal portion of the inguinal ligament which lies medially and also note that there is nothing to prevent the tip of your finger from entering the abdominal cavity (not the peritoneal cavity). Now locate the **femoral nerve** (L2, 3, 4; posterior divisions) which lies outside and lateral to the femoral sheath. After passing beneath the inguinal ligament the femoral nerve breaks up into its terminal branches. Trace these branches and note that the **motor branches** supply all the muscles of the anterior compartment of the thigh.

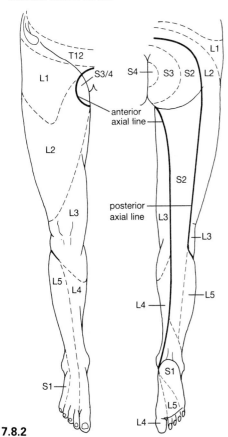

7.8.2

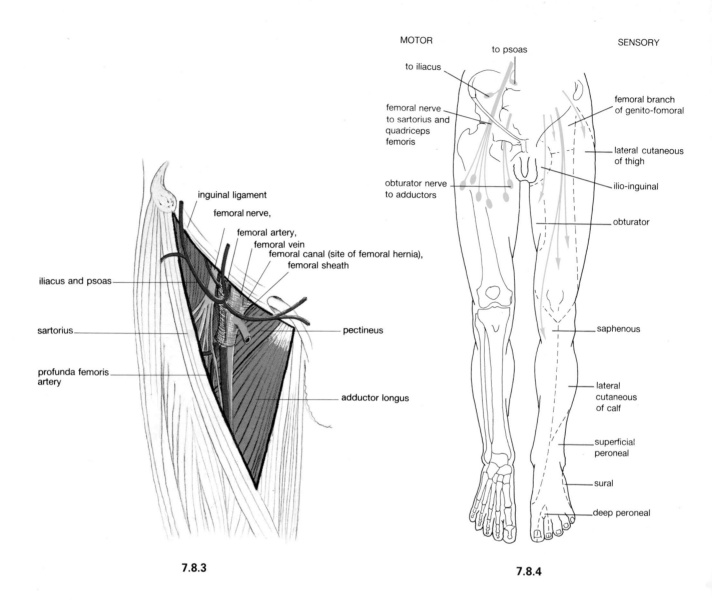

7.8.3

7.8.4

Qu. 8B *If you stretch the quadriceps muscle by tapping the patellar tendon with a patellar hammer you will elicit a 'knee jerk'.*
What are the neural pathways involved in this simple tendon reflex and what different types of information are you eliciting when you perform such a test?

Sensory branches supply the skin over the front and medial aspects of the thigh while a long sensory branch, the **saphenous nerve**, enters the adductor canal and then, accompanying the long saphenous vein, supplies a strip of skin which runs down the medial aspect of the lower limb to the 1st metatarso–phalangeal joint. To trace all these branches you will find it easier to cut across the sartorius (thereby exposing the adductor canal) and reflect it. The femoral nerve also sends branches to the hip and knee joint.
    The **lateral cutaneous nerve of thigh** arises separately from the lumbar plexus, passes under the lateral end of the inguinal ligament, and supplies the skin of the lateral aspect of the thigh.

b. The *obturator nerve* (7.8.4)

Cut through the belly of adductor longus half way along its length and note that adductor brevis lies deep to it. The anterior branch of the obturator nerve lies on the adductor brevis and should be traced from the obturator externus, through which it has passed when leaving the pelvis to enter the thigh. Note the posterior branch of the obturator nerve which lies behind adductor brevis and therefore lies on adductor magnus. The obturator nerve (L2, 3, 4; anterior divisions) supplies all the muscles of the adductor compartment and the skin over it. It also supplies the hip and knee joints.
    It is worth remembering that the ilio–inguinal nerve (L1) also supplies the medial side of the thigh and that the cremasteric reflex (see Vol 2) can be elicited by stroking this area of skin.

Requirements:
    Articulated skeleton
    Prosections of arterior and medial aspects of
        thigh showing the femoral and obturator
        nerves and their branches.

Seminar 9

Innervation of the lower limb: sciatic nerve

The **aim** of the seminar is to trace by dissection the course of the sciatic nerve and its branches.

A. Dissection

a. The *sciatic nerve*
(7.8.1, 7.9.1–7.9.4)

Turn the cadaver on to its front. Cut through the gluteus maximus close to its origin from the ilium and sacrum and reflect it laterally. To do so you will have to sever the muscle fibres which take origin from the sacro–tuberous ligament. Identify gluteus medius and piriformis. Locate the very large sciatic nerve as it emerges from the pelvis into the gluteal region below piriformis. Note that the **posterior cutaneous nerve of the thigh** lies superficial to it and supplies an area of skin that its name suggests. Trace the sciatic nerve as it runs downwards and under cover of biceps femoris to enter the posterior compartment of the thigh. Note that it is separated from the hip joint only by the small muscles which are lateral rotators of the hip.

The **sciatic nerve** (L4, 5, S1, 2, 3) supplies all the muscles of the posterior compartment of the thigh and usually ends by dividing halfway down the posterior compartment of the thigh into the **tibial and common peroneal nerves**. In this position the nerves are lying on adductor magnus and between semimembranosus and semitendinosus medially, and biceps femoris laterally.

Trace the **tibial nerve** as it passes straight downwards behind the knee joint (i.e. through the popliteal fossa) into the leg, noting its close proximity to the posterior aspect of the popliteal artery and vein. The tibial nerve supplies all the flexor muscles of the back of the leg and the sole of the foot; the skin of the lower half of the back of the leg and the lateral side and sole of the foot and the joints of the knee, ankle, and foot.

Cut across the tendo–calcaneus and reflect the superficial muscles of the calf upwards. Follow the tibial nerve from the popliteal fossa into the leg where it lies between the superficial and deep muscles and note that in this position it supplies all the deep muscles. Trace the nerve to the back of the medial malleolus and note its position with respect to the tendons and blood vessels, all of which are passing into the sole of the foot. Immediately behind the malleolus lies the tendon of tibialis posterior and next to it the tendon of flexor digitorum longus, then the neurovascular bundle and most posteriorly the tendon of flexor hallucis longus which grooves the talus as it enters the foot. Here the tibial nerve divides into its **medial and lateral plantar branches (7.9.3)** which behave in

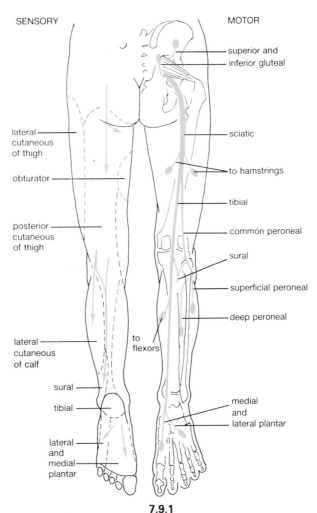

7.9.1

a similar fashion to the median and ulnar nerves of the palm of the hand.

Now return to the sciatic nerve and trace its other terminal branch, the **common peroneal nerve (7.9.4)**. This nerve passes through the popliteal fossa medial to the tendon of the biceps femoris and when it reaches the lateral aspect of the fossa it curves forwards over the lateral aspect of the neck of the fibula. In this position it is superficial and palpable, and is therefore liable to injury as a result of direct trauma or pressure due to a badly applied plaster cast. The nerve supplies branches to the knee joint and to the skin on the back and

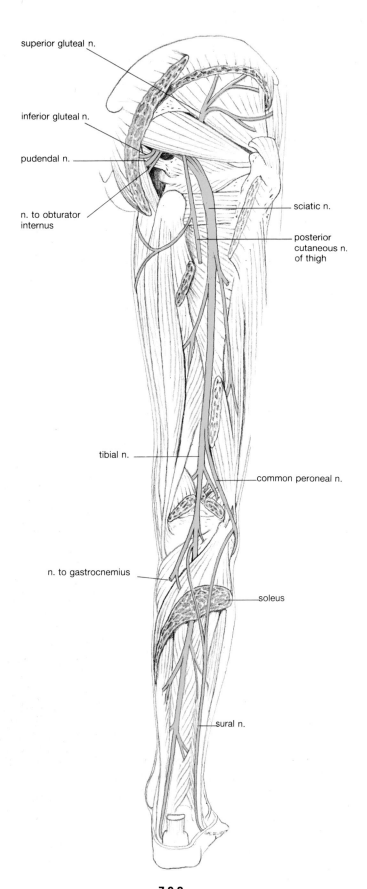

superior gluteal n.

inferior gluteal n.

pudendal n.

n. to obturator
internus

sciatic n.

posterior
cutaneous n.
of thigh

tibial n.

common peroneal n.

n. to gastrocnemius

soleus

sural n.

7.9.2

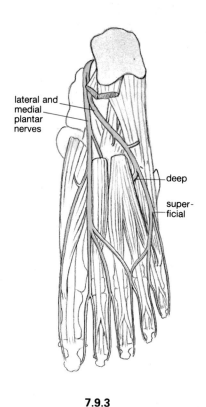

lateral and
medial
plantar
nerves

deep

super-
ficial

7.9.3

postero–lateral aspects of the calf it then passes under cover of fibres of the peroneus longus and divides into the **superficial and deep peroneal nerves**.

The **superficial peroneal nerve** should be followed into the lateral compartment of the leg where it supplies peroneus longus and brevis. About two thirds of the way down the lateral aspect of the leg the nerve becomes superficial and descends over the front of the leg and ankle dividing into branches which supply the skin over the lower lateral aspect of the leg, the front of the ankle and the dorsum of the foot (with the exception of the skin between the big and 2nd toe). Now return to the **deep peroneal nerve**. This nerve enters the anterior compartment of the leg and descends with the anterior tibial vessels which have passed from the posterior compartment of the leg over the interosseous membrane to reach the anterior compartment. As the neurovascular bundle crosses the ankle it can be located halfway between the two malleoli, i.e. between the tendons of extensor hallucis longus medially and extensor digitorum longus laterally. The deep peroneal nerve supplies all the muscles of the anterior compartment and the extensor digitorum brevis on the dorsum of the foot as well as joints of ankle and the foot. It ends by running forward to supply the skin of the cleft between the big and second toe. A sural nerve is formed by branches of both tibial and common peroneal nerves. It runs down the back of the calf with the short saphenous vein and supplies the skin of the back of the calf, heel, and lateral side of the foot. This is the nerve most commonly used by surgeons for grafting a gap between the ends of an injured nerve.

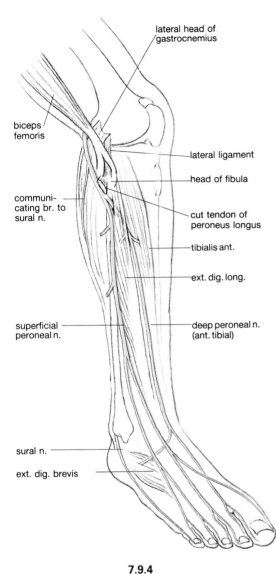

biceps femoris

lateral head of gastrocnemius

communi-cating br. to sural n.

lateral ligament

head of fibula

cut tendon of peroneus longus

tibialis ant.

ext. dig. long.

superficial peroneal n.

deep peroneal n. (ant. tibial)

sural n.

ext. dig. brevis

7.9.4

Qu. 9A *A cyclist received a glancing blow on the side of the leg from a passing car which severed the common peroneal nerve. What might be the results of this injury?*

Qu. 9B *If a patient sustained a gunshot wound which severed the left sciatic nerve in the gluteal region what would be the result of such an injury?*

The effects in the leg of spinal nerve damage

The spinal nerves, or nerve roots, which make up the femoral, obturator, and sciatic nerve, pass from the vertebral canal through the intervertebral foramina. These nerve roots can be individually involved in pathological processes and it is often important to try to differentiate, by symptoms and clinical signs, which roots are involved. In fact the most commonly affected roots are L4, L5 and S1.

The fourth lumbar root
Motor signs. It is not possible by clinical examina-tion to detect wasting or weakness of all the muscles which are partially supplied by L4, but an obvious one is the quadriceps femoris, which receives a large contribution from this root. Some wasting of the quadriceps may be apparent. Also, tibialis anterior and tibialis posterior both receive a significant contribution from L4, so that there may be some detectable weakness of inversion of the foot.

Sensory signs. Some blunting of sensation over the antero–medial aspect of the shin may occur as far as the 'bunion' area of the foot (the metatarso–phalangeal joint of the big toe).

Reflex jerks. Because of the effect on the quadri-ceps muscle there may be loss of the knee jerk. The ankle is not affected.

The fifth lumbar root
Motor signs. Slight weakness of dorsiflexion of foot and extension of the big toe at the metatarso–phalangeal joint may occur. The extensor hallucis longus seems especially vulnerable to loss of function of this root. There may also be a little weakness of eversion.

Sensory signs. Blunting of the sensation may be found extending down the antero–lateral aspect of the shin across the dorsum of the foot to the hallux beyond the first metatarso–phalangeal joint; also the sole of the foot.

Reflex jerks. Neither the knee jerk nor the ankle jerk are affected.

The first sacral root
Motor signs. Some loss of power of the gluteus maximus and of plantar flexion of the foot may be detectable.

Sensory signs. Blunting of sensation extending down the postero–lateral aspect of the shin to the dorsal and plantar surfaces of the outer aspect of the foot may occur.

Reflex jerks. The knee jerk is unaffected, but the ankle jerk may be lost.

Referred pain in the leg
The hip joint is supplied by branches of the femoral, obturator, and superior gluteal nerves, and from the nerve to quadratus femoris. The root values of these nerves extend from L2 to S1. The knee joint is supplied by the femoral, obturator, tibial, and common peroneal nerves. These root values extend from L2 to S3. There is thus considerable overlap of the sensory supply of both joints. This is shown clinically in the presentation of disease; it is a common experience to hear a patient complain of pain down the front of the thigh and knee, only to find that disease lies in the hip joint. For instance, early slipping of the upper femoral epiphysis often causes discomfort in the knee. It is also worth remembering that disease within the pelvis, affecting the sciatic nerve, can cause pain referred down the back of the thigh.

Requirements:
   Articulated skeleton
   Prosections of the lower limb, the sciatic nerve
      and its branches; common peroneal, superfi-
      cial and deep peroneal, tibial, medial, and
      lateral plantar nerves.

CHAPTER 8

The spinal column
Introduction

Bipedalism in hominids emerged about two million years ago as a result of a combination of selection pressures concerned with object carrying, use of tools and weapons, and the need for vision over an increased distance. When Homo sapiens appeared the multiunit flexible spine had already evolved, with the development of specialized joints situated between the spine and skull which enabled the head to rotate thereby increasing the horizontal range of visual scanning. In addition, a series of alternating curvatures of the spine (forward facing in the neck and lumbar regions, and backward facing in the chest and pelvis) also appeared, ensuring the efficient balance of the head on the neck and of the trunk on the lower limbs, and supporting the thoracic cage and shoulder girdle.

The erect posture exposed more of the vulnerable abdomen to danger. This disadvantage was minimised by a reduction in the number of lumbar vertebrae, bringing the ribs closer to the pelvis and thereby increasing protection. Maintenance of the erect posture also meant that the weight of a large part of the body had to be transmitted through the lumbar spine and sacrum to the pelvis and lower limbs; the forces involved being magnified many times in running and jumping. Although the spinal column has become well adapted for these functions in many ways, nevertheless, back problems are relatively common in modern man, occurring more frequently in the lower lumbar spine than in any other region.

Seminar 1

Spinal column

The **aim** of this seminar is to study the spine, which constitutes the central axis of the body and, by its form, dominates the appearance of the trunk.

A. Living anatomy

The spine consists of a series of vertebrae, ligaments, and intervertebral discs. This flexible column of bone, ligament, and muscle encloses a canal and protects the spinal cord within it. It supports the weight of the trunk and transfers the weight to the pelvis; the discs provide shock–absorbing resilience. It can be moved by a series of postural and 'prime mover' muscles to orient the limb girdles and head; it also provides attachment for many other muscles (e.g. girdle muscles) and for the ribs. The vertebral bodies have an additional function since they contain red bone marrow throughout life.

Compare your partner's back (**8.1.1**), with the back of an articulated skeleton (**8.1.2**) and with radiographs of appropriate regions of the spine. Note that the spinal column normally lies strictly in the median sagittal plane (provided that the legs are of equal length).

Alterations of the normal curvature may occur (**8.1.3**, **8.1.4**). Any deviation to the right or left of the midline vertical constitutes a deformity known as a **scoliosis** (**8.1.4a**). Minor degrees of 'postural' scoliosis are unimportant and may disappear on flexion of the spine. 'Structural' scoliosis is a permanent deformity. Scoliosis may result from maldevelopment (a hemivertebra, failure of one side of a vertebra to develop), or from contracture of one side of the chest wall due to underlying lung disease. Most scolioses, however, are of unknown origin and involve rotational displacement of all the adjacent vertebrae, the bodies of the vertebrae pointing to the concavity of the curvature. The normal **kyphosis** may be increased (**8.4.1b**), particularly if the bones become weak through lack of calcium as may occur in the elderly. A severe scoliosis may be associated with an increased kyphosis which severely diminishes the volume of the thoracic cavity and causes lung and heart problems. A localized injury or collapse of a single vertebral body can produce an angular kyphosis, the visible bony apex of which is called the kyphus.

Identification of vertebral levels on the back

Run your finger down the midline of your partners' back, starting at the base of the skull. Upper cervical spinous processes lie deeply within the neck muscles; the first easily felt spinous process (the 'vertebra prominens') is usually C7 or T1.

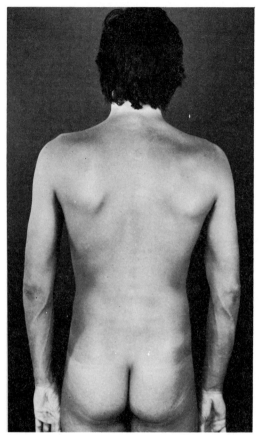

8.1.1

Draw a line between the uppermost parts of the iliac crests and note that it extends between the spinous processes of L4 and L5. The scapula is very mobile but, with the arms at rest by the side, the spine of T3 is usually at the level of the base of the spine of the scapula and the spine of T7 usually lies at the level of the lower angle of the scapula.

Qu. 1A *Locate the spinous process of L1. The spinal cord ends at this level in the adult although it extends to L5 in the new born child. Why is this?*

Feel the prominence of muscular columns (erector spinae) which lie on each side of the lumbar vertebral spines.

Qu. 1B *When your partner stands on one leg, these become particularly prominent on one side. On which side, and what is the explanation?*

Note also the dimples that overlie the posterior superior iliac spines of the pelvis. Next, assess the **range of movement of the spine**:

a. Flexion
Ask your partner to stand, knees straight, and then bend forwards to try to touch the ground with the tips of his fingers. The distance between fingertip and ground provides a rough measure of the flexion

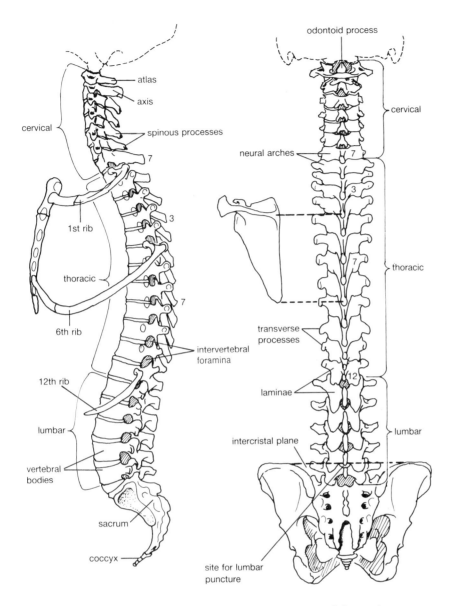

8.1.2

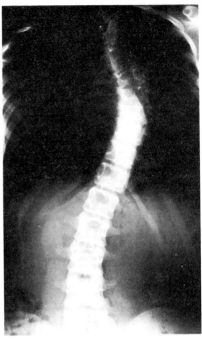

8.1.3

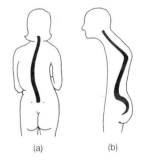

(a)   (b)

8.1.4

achieved. As a true measurement it is complicated by the degree of hip flexion and the tension of the hamstring muscles.

Ask your partner to stand erect and measure the distance between the spine of S1 and the vertebra prominens. Repeat this with your partner flexed i.e. touching his toes. Comment on the results.

S1 to vertebra prominens in erect posture ☐ cm
in full flexion ☐ cm

b. Lateral flexion
This is assessed by recording the level attained by the tips of the fingers reaching down the side of the thigh to the knee, without rotation of the trunk or flexion of the knees.

c. Extension and rotation
These are less easy to measure and are usually assessed by eye.
From examinations of your partner assess which regions of the spine contribute most to

a) flexion and extension —
b)   rotation —
c)   lateral flexion —

Development of the spine (8.1.5)
The first axial element to be formed is the **notochord** with its sheath lying between the neural plate/tube and the dorsal wall of the yolk sac. Cells of the **sclerotome** (part of the mesodermal somites) proliferate and migrate around the notochord and around the neural tube to form a mesenchymal precursor of the vertebral column.

The mesenchyme of the sclerotomes, which is segmentally arranged, becomes divided by a fissure. This marks the position in which the intervertebral disc will ultimately develop. Above and below the fissure, a perichordal disc contributes to both the intervertebral disc and the centra. The primordia of the centra are thus formed from the adjoining parts of two adjacent sclerotomes and are intersegmental. The segmental spinal nerves therefore pass out between individual vertebrae; the intersegmental arteries (e.g. the intercostal arteries, lumbar arteries) lie on the vertebral bodies.

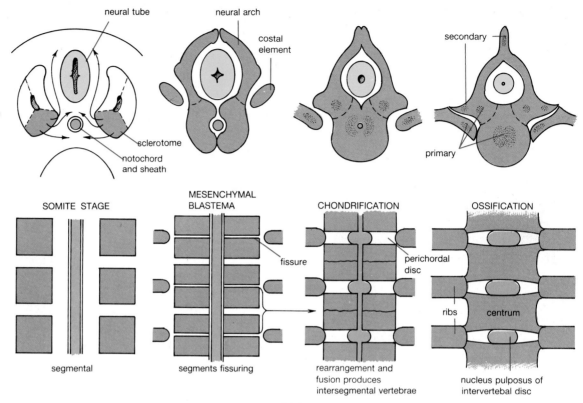

8.1.5

Chondrification of the centra starts at about the 6th week *in utero*, from two centres which rapidly fuse. Each half of the neural arch chondrifies from a centre at its base. Costal elements chondrify separately. The neural arch chondrification does not unite in the midline until the 4th month *in utero*.

Ossification of typical vertebrae from **primary centres** starts at 7–8 weeks *in utero*. The ossification centres are in the same position as the chondrification centres. The neural arches unite as bone during the first year, but the neurocentral joint between the arch and the centrum persists for the first few years of life. At puberty **secondary centres** appear in the spinous process and transverse process, and two circumferential, annular epiphysial discs form at the cranial and caudal ends of the vertebral bodies.

Particular ossification patterns occur in the atlas, axis, and lumbar vertebrae. The anterior arch of the atlas is formed from the **hypochordal bow**—tissue which unites the costal elements anterior to the centra. The centrum of the 1st cervical somite forms the odontoid peg of the axis. The pattern in the sacrum is essentially similar to that in typical vertebrae, except that the vertebrae fuse with each other and with the costal elements which form the alae.

Stand to one side and note the three spinal curvatures: **cervical lordosis** (dorsalward concavity); **thoracic kyphosis** (dorsalward convexity) and **lumbar lordosis**. Compare the spinal curvature of different subjects, particularly the degree of the thoracic kyphosis. One way to assess this roughly and quickly is to ask the subject to stand with his thoracic spine pressed against a wall, holding the head in the **natural erect posture**. The degree of kyphosis can be assessed by the horizontal distance from the wall to the tragus of the ear (the cartilaginous lump just in front of the meatus).

What is the wall–to–tragus distance in
a) your partner   ☐ cm?
b) yourself   ☐ cm?

Congenital anomalies of the spine

Congenital anomalies of the spine are relatively common. The spine is a functional unit made up of many vertebrae each of which is formed from a number of elements which must segment, chondrify and ossify correctly. A defect in a single vertebra such as a hemivertebra arrows (**8.1.6**) can so disturb the normal anatomy as to make the spine **unstable** so that when the stresses of post–natal life are imposed, progressive deformation will result. Other anomalies such as fusion of two or more adjacent vertebral bodies may produce a minor loss of movement which can easily be compensated for, but they are essentially **stable** and do not lead to progressive deformity. Many spinal anomalies are associated either primarily or secondarily with nervous system defects and damage. **8.1.7** is a radiograph of a young patient with spina bifida in which the absence of lumbar neural arches (arrows) can be seen. When considering specific deformities think of their effects not only on the mechanics of the spine but on the thoracic cavity and respiration, and on the spinal cord and nerves.

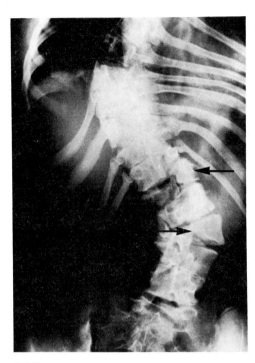

8.1.6

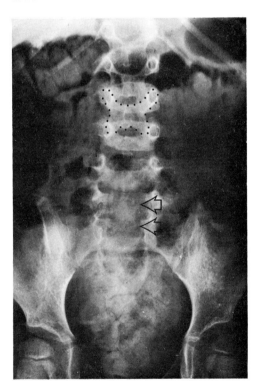

8.1.7

The bony skeleton (8.1.2, 8.1.8)

Examine vertebrae from different regions of the spine (**8.1.8a, b**)

On each vertebra you should identify:

- the **centrum** or **body** of the vertebra

- the **vertebral arch**. Each arch comprises, on each side, a **pedicle** attached to the vertebral body and a broad **lamina**. Together with the vertebral

body, the arch encloses the **vertebral foramen**.

- a **spinous process** which projects from the arch dorsally in the midline

- two **transverse processes**

- paired **superior and inferior articular facets**, which project from the pedicles so that the vertebral arch articulates with the vertebra above and below.

Note the shape and orientation of the superior and inferior articular facets in the different regions. These form synovial joints at which a small amount of gliding movement occurs.

Qu. 1C *How does the shape of the articular surfaces relate to the movement possible in a particular region?*

Multiple small vascular foramina can be found on the bone. The largest are seen on the dorsal surface of the centrum where basivertebral veins drain the red marrow to the internal vertebral venous plexus.

Costal elements develop in all regions but normally form separate ribs only in the thoracic region. In the cervical region they form the part of the transverse process anterior to the foramen transversarium; in the lumbar region they are incorporated into the transverse process. Abnormally separate ribs may develop in either region but are most common at C7 and L1.

Qu. 1D *Would you expect a greater proportion of people with cervical or lumbar accessory ribs to complain of symptoms, and why?*

Cervical vertebrae:

C1 The atlas: The atlas does not have a centrum, during development this became the odontoid process of the axis. Its short **anterior arch** has an anterior tubercle to which the anterior longitudinal spinal ligament is attached at its apex, and an articular facet for the dens. The larger **posterior arch** is grooved on its upper surface by the vertebral artery and a posterior tubercle represents the spinous process. The **lateral masses** are each pierced by a foramen transversarium for the vertebral artery. The attachment of the transverse ligament of the atlas can be seen in the vertebral canal and the **transverse processes** are prominent. Large superior facets articulate with the occipital condyles.

C2 The axis: The axis is easily distinguished by the **odontoid peg (dens)** which bears articular facets for the anterior arch of the atlas on its anterior surface, for the transverse ligament of the atlas on its posterior surface, and for the attachment of the alar ligaments above. The **body** has prominent oval articular facets above which rotation on the atlas occurs. The **laminae** are very thick and the **spinous process** large and bifid as in most other cervical vertebrae.

Cervical vertebrae C3–6 are the 'typical' cervical vertebrae. Each has a relatively small **body**, a **transverse process** pierced by a foramen transversarium with an anterior tubercle (costal element) and posterior tubercle. Note the lips on the

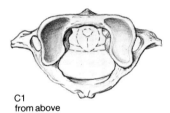

C1
from above

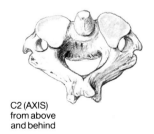

C2 (AXIS)
from above
and behind

8.1.8a

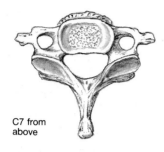

C7 from
above

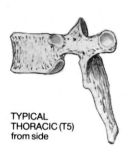

TYPICAL
THORACIC (T5)
from side

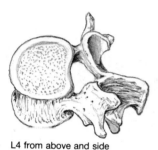

L4 from above and side

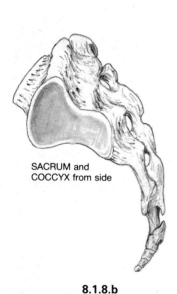

SACRUM and
COCCYX from side

8.1.8.b

upper surfaces of the bodies and the shape and orientation of the articular facets. The **spinous processes** are bifid for the attachment of the ligamentum nuchae.

Cervical 7 vertebra is the 'vertebra prominens' —the first that can easily be felt as the examining hand descends the neck. Its spinous process is horizontal and not bifid, and the vertebral artery does not traverse its (often narrow) foramen transversarium.

Thoracic vertebrae have bodies of increasing size which typically bear upper and lower hemifacets for the heads of the ribs. The vertebral foramen is relatively small; the laminae are thick, broad and overlapping; the spinous processes are long and slant downwards and backwards; the transverse processes are substantial and most bear facets for articulation with the tubercles of their numerically corresponding ribs.

T1 has a complete upper facet on the body and a thick, horizontal spine
T9 often fails to articulate with the 10th rib
T10 articulates with only the 10th rib
T11 articulates with only the 11th rib and lacks facets on its transverse processes
T12 articulates with only the 12th rib and lacks facets on the transverse processes.
Interlocking articular facets of the lumbar type start at the lower border of T11 or T12.

Lumbar vertebrae have large, deep bodies; short, stout pedicles; a quadrangular spinous process and long thin transverse processes (costal elements); the articular processes are of the interlocking type. L5 has a short, thick, conical transverse process that gives attachment to the ilio–lumbar ligament which connects it to the pelvis.

The sacrum is large, triangular, and normally consists of 5 fused vertebrae. The base articulates with L5 at the sacrovertebral angle. Anteriorly it projects as the sacral promontory. The pelvic surface is marked by 4 anterior sacral foramina. Lateral to the foramina the costal elements are fused to form the lateral part or ala (wing) of the sacrum. Dorsally the median sacral crest represents the spinous processes and 4 dorsal foramina can be seen. Laterally the fused transverse and costal elements form the L–shaped articular surface in its upper part. The whole sacrum forms a gently curving roof and posterior wall of the pelvis. Below it the coccyx consists of 3–5 rudimentary vertebrae.

The **number of vertebrae** in different regions may vary, usually by only one element. Most commonly this occurs in the lumbosacral region, due to the complete or incomplete sacralization of a lumbar vertebra.

Examine in particular the articulation between L5 and the sacrum. The upper surface of the body of S1 slopes downward and forward when the trunk is erect. Forward movement of L5 on S1 is normally

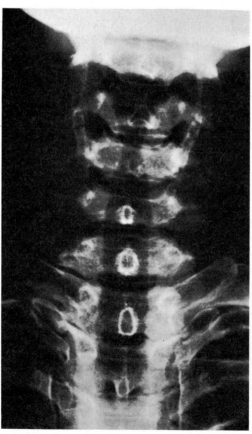

8.1.9

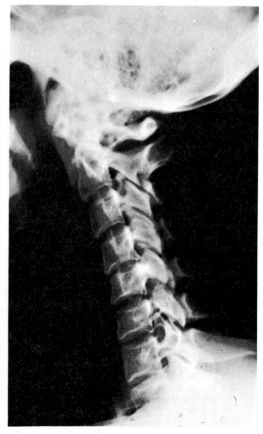

8.1.10

prevented by the articular facets of the sacrum which lie anterior to those of L5 and by the integrity of the intervertebral disc between the vertebrae.

There is a condition known as spondylolisthesis, which may begin as an infant starts to toddle, in which a defect occurs between the superior and inferior facets of L5, a region known as the pars interarticularis. The body of L5, carrying the spinal column above it, can then move forward, leaving the posterior facets, laminae, and spinous process behind. As displacement of the spine continues, compression of the nerves in the sacral canal (particularly S1) can occur.

B. Radiographs

Now examine the radiographs of the different regions of the spine (**8.1.9–8.1.16**) and identify all the components you have located on the dry bones. An oblique view of the lumbar spine (**8.1.17**) presents a 'Scottie dog' appearance in which the dog's nose is the transverse process, its ear the superior articular facet, and its eye the pedicle. The collar is the 'pars interarticularis'—the anterior part of the lamina.

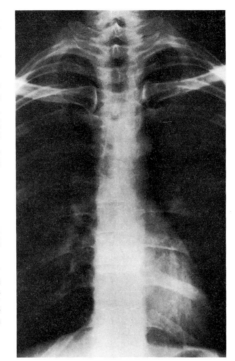

8.1.12

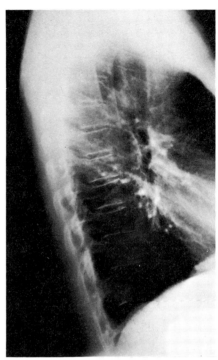

8.1.13

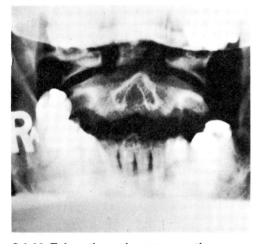

8.1.11 Taken through open mouth to show the axis

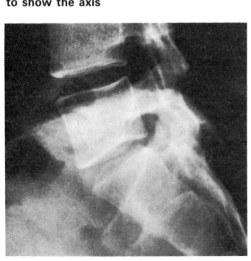

8.1.14

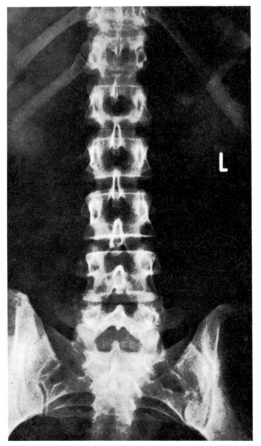

8.1.15

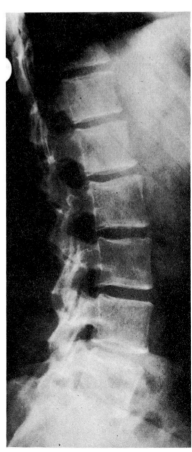

8.1.16

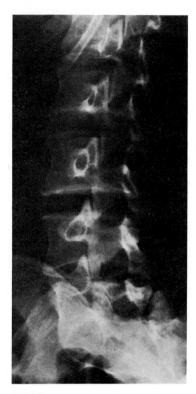

8.1.17

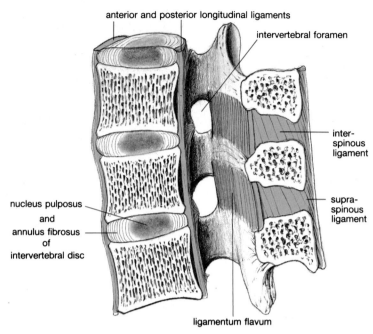

anterior and posterior longitudinal ligaments

intervertebral foramen

inter-spinous ligament

supra-spinous ligament

nucleus pulposus and annulus fibrosus of intervertebral disc

ligamentum flavum

8.1.18

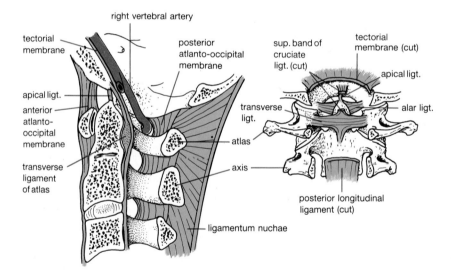

right vertebral artery

tectorial membrane

apical ligt.

anterior atlanto-occipital membrane

transverse ligament of atlas

posterior atlanto-occipital membrane

sup. band of cruciate ligt. (cut)

tectorial membrane (cut)

apical ligt.

transverse ligt.

alar ligt.

atlas

axis

posterior longitudinal ligament (cut)

ligamentum nuchae

8.1.19

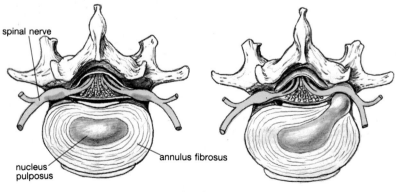

spinal nerve

nucleus pulposus

annulus fibrosus

8.1.20

C. Prosections

The ligaments of the spine (8.1.18)

The individual vertebrae are held together by a series of **long** and **short** ligaments.
On prosections identify and examine:

Short:

interspinous ligaments between adjacent spinous processes,

intertransverse ligaments between adjacent transverse processes,

ligamenta flava between the laminae of the neural arches; these are yellow, elastic ligaments.

Long:

anterior longitudinal ligament on the anterior aspect of the centra,

posterior longitudinal ligament on the posterior aspect of the centra,

supraspinous ligament over the tips of the spinous processes. This becomes the **ligamentum nuchae** between C7 and the skull.

In the region immediately beneath the skull, both vertebrae and ligaments are specialized in relation to the movements that occur between the occiput and the atlas, and between the axis and the atlas. Move your own head and study the bones to discover the movements of this region. Nodding flexion/extension occurs at the atlanto–occipital joint with very minor lateral flexion possible. The atlanto–axial joint is specialized to allow rotation; other movements being prevented by the restriction of the odontoid peg within the anterior arch of the atlas by the transverse ligament.
On a prosection (**8.1.19**) identify the:

anterior atlanto–occipital membrane—continuous with the anterior longitudinal ligament.

tectorial membrane—continuous with the posterior longitudinal ligament.

posterior atlanto–occipital membrane—the homologue of ligamenta flava.

cruciate ligament—this comprises the **transverse ligament of the atlas** with extensions passing upward and downward.

alar ligaments which extend from the top and sides of the odontoid process laterally to the margin of the foramen magnum.

apical ligament—this thin band, formed from the notochord, may have a small bone, a pro–atlas, within it.

Qu. 1E *What is the function of the transverse ligament of the atlas?*

Qu. 1F *What movements do the alar ligaments restrain?*

Intervertebral discs (8.1.20)

Examine the dissected intervertebral discs provided. Identify the fibrous outer ring, the **annulus fibrosus**. Although it may be difficult to see them, the annulus is composed of a series of layers, and adjacent layers of the fibres pass obliquely in different directions giving great torsional strength. Identify the **nucleus pulposus** lying within the annulus. In the child the annulus is thin and the nucleus central. With ageing the annulus becomes thicker at the front and the nucleus comes to lie more posteriorly in the disc. For this reason any herniation (protrusion) of the nucleus pulposus is likely to occur in a posterior or posterolateral direction.

Intrinsic muscles of the spine

There are essentially three groups of intrinsic spinal muscles:

a) **flexors (8.1.21)**—this relatively weak anterior group is found in the neck (longus colli) and extends from T4 to the base of the skull (longus capitis) with attachments at all levels. In the lumbar region psoas major acts as a flexor. The flexors are supplied segmentally by anterior primary rami of spinal nerves. Muscles placed anteriorly in the body but separated from the spine also act as powerful spinal flexors; examples of such muscles are sternomastoid in the neck and rectus abdominis in the trunk.

b) A **lateral** group—the **scalene** muscles. These run from cervical transverse processes to ribs, produce lateral flexion of the neck, and act as 'guy lines' supporting the head on the trunk (see Vol. 3).

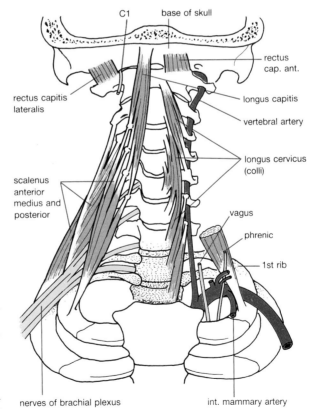

8.1.21

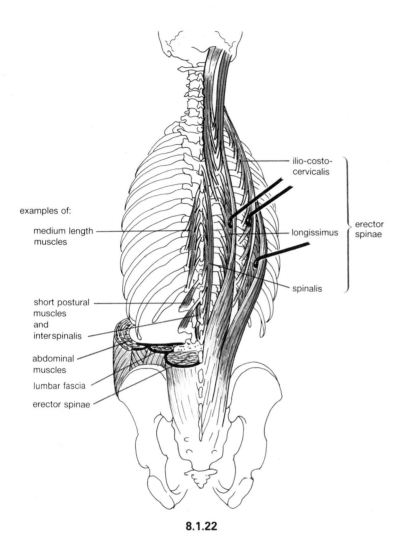

examples of:

medium length muscles

short postural muscles and interspinalis

abdominal muscles

lumbar fascia

erector spinae

ilio-costo-cervicalis

longissimus

spinalis

} erector spinae

8.1.22

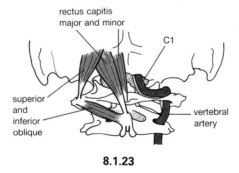

rectus capitis major and minor

C1

superior and inferior oblique

vertebral artery

8.1.23

c) **Postural muscles and extensors (8.1.22)**—this large group of muscles lies posteriorly and is arranged as a series of long, medium, and short muscles. The shortest muscles, i.e. those between individual transverse processes and between individual spinous processes, lie most deeply and are mainly **postural** in function, producing only a little extension and rotation. Medium length muscles lie superficial to these and run from transverse processes to spinous processes of vertebrae adjacent or of more distant regions. The longest and most superficial muscles form the **erector spinae**. This group is firmly anchored to the dorsum of the sacrum and is the prime mover in extension of the spine. Fibre bundles are attached to spinous processes, transverse processes, and costal elements. Some of its components extend to the occiput and thus produce extension of the head. In the region immediately beneath the skull, a specialized group of **suboccipital** muscles **(8.1.23)** controls extension and rotation at the atlanto–occipital and atlanto–axial joints respectively. These will be examined with the head and neck.

All the muscles of the extensor group are supplied by the **posterior primary rami** of the spinal nerves which also supply the skin of the back.

Contents of the spinal canal (8.1.24)

Within the **spinal canal** lie the spinal cord and spinal nerve roots, surrounded by protective membranes, the **meninges**. The outer tough fibrous layer is called the **dura mater** and this extends from the head throughout the length of the spine to S2 where it narrows to a thin cord. In the **extradural space** between the bone and dura, lies an **internal vertebral venous plexus** supported by a little fat. This plexus drains the marrow in the vertebral bodies via basi–vertebral veins which emerge from the dorsal surfaces of the centra, and it communicates with a plexus outside the spinal column, the **external vertebral venous plexus**, via the intervertebral foramina. Blood flow in this system offers a venous pathway to the heart which can by–pass the abdominal cavity, particularly when intra–abdominal pressure is raised. It is also an important route for the spread of (pelvic) malignancy since it is without valves and is connected to veins of the pelvis, the flow of blood eddying to and fro through the external and internal plexuses with, for example, changes of posture.

The dura mater is lined by a cobwebby membrane, the **arachnoid mater** within and by which the spinal cord is suspended in **cerebrospinal fluid** (CSF). The CSF–filled **subarachnoid space** ends at S2 level.

Samples of CSF can be obtained by **lumbar puncture**, and submitted to cytological and chemical study. Before lumbar puncture, the pressure of the CSF in the cranial cavity must be assessed by examining the optic disc of the eye (see Vol. 3). The skin is anaesthetized and the patient lies on his side, curled up as much as possible.

Qu. 1G *Why is a patient undergoing lumbar puncture asked to curl up as much as possible?*

The lumbar puncture needle is introduced in the midline, pointing forward and slightly toward the head, between the spines of L4 and L5 vertebrae (i.e. the level of the intercristal plane, see *Abdomen*), through the interspinous ligament, through the ligamentum flavum, and so into the spinal canal. Advancing further, the needle will pierce the dura and arachnoid membranes and enter the CSF–containing subarachnoid space which, below the level of L1, contains the mass of lower lumbar and spinal nerve roots known as the **cauda equina**. Before any fluid is withdrawn the pressure is measured with a manometer. Examine the specimen of lumbar spine and the needle provided; pass it as you would in a lumbar puncture mentally noting the stuctures which are transvered.

This access to the spinal dural sac also permits:
1. an anaesthetist to inject local anaesthetic into the CSF around the roots of the cauda equina, to allow surgery to be performed on the pelvis (e.g. during child birth) or legs. Such **spinal anaesthesia** is especially useful where general anaesthesia is contra–indicated. Extradural application of anaesthetic around emerging spinal nerves (**epidural** anaesthesia) can also be used.

Qu. 1H *When a needle is in the extradural space for injection of fluid, can any CSF be withdrawn?*

2. a radiologist to inject non–irritant radio–opaque dye into the cerebrospinal fluid. Radiologic examination (a myelogram) (**8.1.25**) then reveals the shape of the dural sac, and the outline of the dural 'sleeves' at the origin of the roots as they pass through the dura. The patient is tilted to allow the viscous dye to reach the appropriate level. A distortion of the normal outline of the dural sac can reveal, for instance, protrusion of an intervertebral disc (arrow).

The spinal cord and spinal nerve roots are usually well protected by the vertebral bodies and by the joints and ligaments which connect them. However, in severe injuries (a car accident, fall from a horse, roof fall in a mine etc.) the spine may break, especially at the junctions of relatively mobile and immobile segments (the lower cervical and thoracolumbar regions). One vertebra and the spinal column above it can shear away completely from the adjacent vertebra and spinal column below. It is important for the doctor attending such a patient to appreciate the damage that can be caused by movement after, as well as at the time of, the accident.

Qu. 1I *If the injury was between C6 and C7, such that the spinal cord was transected, what would be the results?*
Would the patient be able to breathe?
What arm movements would he still be able to perform?

If the injury were of a similar nature but between T12 and L1, there could be damage to either or both the spinal cord (its terminal sacral part) and also to the nerve roots that pass down to their respective intervertebral foramina. It is important to distinguish between nerve root and spinal cord

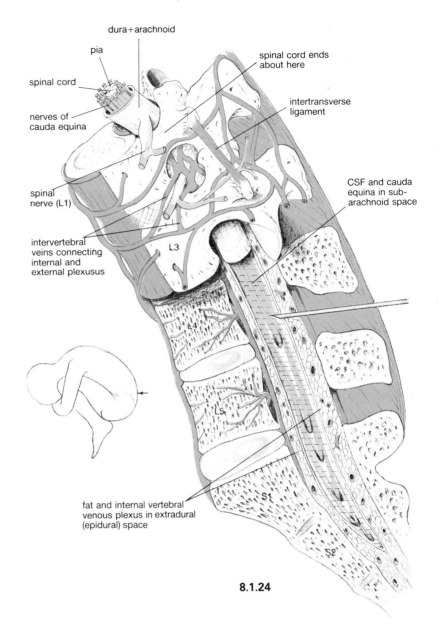

dura+arachnoid
pia
spinal cord
nerves of cauda equina
spinal nerve (L1)
intervertebral veins connecting internal and external plexusus
spinal cord ends about here
intertransverse ligament
CSF and cauda equina in sub-arachnoid space
L3
L4
L5
S1
S2
fat and internal vertebral venous plexus in extradural (epidural) space

8.1.24

damage since, in the former, there is the possibility of some recovery as fibres can regrow down Schwann cell sheaths. With cord injuries the damage will be permanent.

Qu. 1J *How will the results of an injury of this type differ from those of an injury at C6/7?*

If the level of transection is below L1, then only nerve roots will be damaged.

Because the spinal canal is surrounded by bone it follows that any structure within the canal can expand only at the expense of the other contents. An expanding tumour or disc prolapse in the canal can very quickly cause compression of the spinal cord or its roots. One of the most common causes of nerve damage in the spinal canal is the posterior protrusion of part of the nucleus pulposus through a degenerate annulus fibrosus (**8.1.20**). The discs most commonly involved are those between L4/L5 and L5/S1. The protrusions usually occur posteriorly, but to one or other side of the posterior longitudinal ligament.

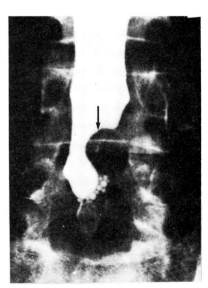

8.1.25

Examine the skeleton and note the relationships of the intervertebral foramina and the anterior and posterior sacral foramina. **8.1.20** shows a transverse view of the lumbar spinal canal. A disc protrusion between L4 and L5 impinging on the L5 nerve root is shown; the lower lumbar region is the most common site of disc protrusion. Very large disc protrusions may affect several nerve roots together.

Qu. 1K *What would be the effect of such a compression of L5 nerve root?*

Disc lesions may occur in the cervical region (and very rarely in the thoracic region). In these regions not only nerve roots but also the anterior aspect of the spinal cord may be compressed. This anterior part of the cord contains motor nerve cell bodies and descending tracts which control them.

Spinal nerves, formed by the union of anterior and posterior nerve roots, pass out of the spinal canal through the intervertebral foramina. Examine the boundaries of an intervertebral foramen: they are formed by edges of the centrum and pedicle, the intervertebral disc, and the facets of the posterior intervertebral joints. Expansion of any of these caused, for example, by posterolateral disc protrusion, arthritic joint changes, or bony overgrowth, will encroach on the intervertebral foramen and damage the spinal nerve within.

Lie your partner supine on a couch and raise one of his legs making sure that the knee is fully extended. There is a limit to which this procedure (the straight leg raise) can be performed without causing discomfort. As the straight leg is raised the spinal nerve roots which contribute to the sciatic nerve are stretched and move several millimetres through their respective intervertebral foramina thereby pulling the dural sac downwards and to one side. If the nerve root is already being compressed this test will increase the pain. Dorsiflexion of the foot during a straight leg raise will have the same effect for similar reasons. The extent to which a 'straight leg raise' can be achieved in comfort can be used to assess progress of a disc lesion (**8.1.26**). A similar test of the femoral nerve can be done by getting the patient to lie in a prone position and flex the knee.

8.1.26

Requirements:

Articulated skeleton; separate vertebrae of the different regions; separate ribs.
Prosections of the spine; ligaments; intervertebral discs; intrinsic muscles.
Prosections of the joints, ligaments, and muscles of the occipito–atlanto–axial region
Prosections of the dorsal aspect of the spine with laminae removed to expose the meningeal coverings of the spinal cord and cauda equina
Radiographs of different regions of the spine

CHAPTER 9

Answers to questions

Upper limb

Seminar 1

1A The principal form of the bone is determined genetically but the extent of ligament and muscle markings is determined by the tensile forces exerted by these tissues on the bone.

1B A clavicle can be 'sided' by recognizing the superior and inferior surfaces, the larger medial end, and the convexity of the anterior border in its medial third (check your own right and left clavicles by feeling the superior and anterior border').

1C The scaphoid, because it has the greatest contact with the forearm bones.

1D **6.1.3** is abnormal. A large radio–opaque excrescence (bone) is seen on the profile of the medial border of the humerus which is consequently not well defined. The outer cortex is absent in the area of irregularity. This is produced by a tumour of bone–producing cells.

1E The medulla or cavity of the bone is wider and the cortex is thinner in the abnormal part of the shaft. This is due to a cyst within the humerus.

1F The first metacarpal has a secondary centre of ossification at its proximal end, as have the phalanges.

Seminar 2

2A The clavicle usually lies uppermost because the most frequent cause of dislocation of the acromio–clavicular joint is a direct blow on the acromion. The coraco–clavicular ligament would be ruptured.

2B A fall on the point of the shoulder, e.g. in a riding accident.

2C The weight of the upper limb and contraction of deltoid fibres would depress the outer end of the clavicle. The relation of the clavicle to the coracoid is maintained by the strong coraco–clavicular ligament. The sternomastoid muscle would tend to elevate the inner fragment through its attachment to its medial end.

2D The rhomboids pull the scapulae towards the midline thus bracing the shoulders.

2E Latissimus dorsi adducts, extends, and medially rotates the humerus.

2F Pectoralis major adducts and medially rotates the humerus at the shoulder joint and, through the action of the clavicular head, helps to flex the joint.

2G Pectoralis minor pulls the scapula (and therefore the upper limb) forwards (protraction) around the chest wall.

2H By tilting the chin laterally and upwards towards the opposite side against resistance, thereby approximating the mastoid process to the sternum. If the muscles on both sides contract the head is pulled forwards (protracted) and downwards.

Seminar 3

3A Yes. Probably many other joints, especially the sternoclavicular. Any movement of the shoulder will require postural adaptation of other parts of the body to maintain balance.

3B No, the arm cannot be raised vertically above the head if the scapula is fixed.

3C Synovial fluid lubricates the joint, distributes pressure within it, and helps to nourish the hyaline articular cartilage.
When pressure on synovial fluid is increased cross–linkages within its glycoprotein molecules decrease and the fluid becomes less viscous, and vice versa (the principle of 'non–drip' paint).

3D Shoulder dislocation usually occurs antero–inferiorly into the dependent, least well–supported part of the capsule.

3E The axillary nerve, since it lies just beneath the dependent part of the capsule.

3F Subscapularis adducts and medially rotates the humerus. In particular it helps stabilize the shoulder joint as part of the rotator cuff.

3G Teres major and the tendon of latissimus dorsi in addition to subscapularis form the posterior axillary wall.

3H Supraspinatus initiates abduction of the arm at the shoulder joint.

3I Deltoid is the 'power' abductor but cannot initiate the movement; serratus anterior and trapezius rotate the scapula around the chest wall to increase the effective movement of the humerus; the rotator cuff muscles stabilize the shoulder joint; as does the long head of biceps.

3J Extension of the humerus.

3K Deltoid abducts the arm.

3L The humerus has become dislocated from the glenoid fossa and the 'hard lump' that can be felt is the head of the humerus.

Seminar 4

4A The elbow joint has been dislocated. The lower end of the humerus has been displaced forward from the trochlear notch of the ulna, and the radio–humeral joint is also dislocated.

4B Brachialis flexes the elbow joint. Coraco-brachialis flexes and adducts the arm at the shoulder joint.

4C Triceps extends the forearm at the elbow joint.

4D Biceps flexes the elbow joint whether the forearm is in the prone or supine. By its attach-ment to the radius it is also a powerful supinator with the forearm in the 'position of function'—i.e. elbow semiflexed.

4E Triceps.

4F Biceps and brachialis, the flexors of the forearm, acting against gravity.

4G Normally, if you grasp the medial and lateral epicondyles of the humerus with your thumb and middle finger from posteriorly, your index finger will rest easily on the tip of the olecranon thus forming the third point of an equilateral triangle. If you cannot do this then the elbow is dislocated. In a supracondylar fracture the triangle would be maintained.

Seminar 5

5A Supination; therefore screws always have right hand threads. Remember that biceps is a powerful supinator in addition to the supinator muscle.

5B Because the wrist joint abducts or adducts to readjust to the central axis of the forearm and hand.

5C The normal axis passes through the middle finger. Note that this entails a small lateral devia-tion of the distal end of the ulna as the radius swings around it in supination.

5D At both the wrist and metacarpo–phalangeal joints, flexion and extension, abduction and ad-duction, and circumduction (being a mixture of the other movements) occur.

5E Medial and lateral rotation cannot occur.

5F The interphalangeal joints are classified as hinge joints because movement at the joints is limited to flexion and extension.

5G The movement of opposition enables Man to achieve a precise or 'precision' grip which is adaptable to a wide variety of differently–shaped/sized objects and which allows fine movements of the fingers as they hold an object.

5H Development of higher neural centres which control motor movement.

Seminar 6

6A Acting as prime movers:
Pronator teres pronates the forearm and, by its humeral attachment, contributes to elbow flexion.

Flexor carpi ulnaris adducts and flexes the hand at the wrist joint.
Flexor carpi radialis flexes and abducts the hand at the wrist joint.
Palmaris longus tenses the palmar aponeurosis and helps to flex the wrist.
Flexor digitorum superficialis flexes the wrist, the metacarpophalangeal joints, and the proximal interphalangeal joints.
All of these muscles contract appropriately in circumduction; and in postural fixation of the wrist to provide a stable support for finger movement.

6B Yes, because biceps also acts to supinate and pronator teres to flex.

6C The hand is extended and abducted at the wrist joint in a 'power grip'. That extension is important is well demonstrated by the ability to break someones grip by flexing their wrist joint.

6D Virtually all: the glass is gripped, the wrist stabilized, and its position delicately adjusted as the elbow and shoulder joints are flexed against gravity.

6E Because if it was, pronation and supination would be limited.

Seminar 7

7A An outer layer of connective tissue containing vessels and nerves (tunica adventitia); a middle layer of smooth muscle and elastic tissue (tunica media); and an inner connective tissue and endo-thelial layer (tunica intima).

7B An artery transmits the pulsations of the blood derived from the contraction of the ventri-cles of the heart (systole).

7C A deep vein or veins is connecting with the superficial vein.

7D First, by direct pressure on the palm of the hand (after first checking that no sharp object is still embedded there) and by elevating the limb. If this fails, then compression of the artery proximal to the injury is required. A tourniquet can be applied to the upper arm, or the axillary or subclavian arteries can be compressed. It is essential that a tourniquet be applied for a minimum period only (about 30 min), or tissue death might occur.

7E To maintain the higher blood pressure in the arteries and smooth the pulse pressure.

7F The arterial obstruction lies at the junction of the right subclavian artery with the common carotid artery.

7G Lymph from the little finger drains along the course of the basilar vein to nodes in the axilla; possibly a gland close to the medial epicondyle would also be enlarged.

Seminar 8

8A Serratus anterior would be incapable of holding the scapula close to the chest wall and of pulling it around the chest wall; forward thrusting movements would be weak and the scapula would 'wing' if they were attempted.

8B Abduction of the humerus at the shoulder joint would be weak (supraspinatus); extension of the shoulder likewise (infraspinatus).

8C To redistribute the motor and sensory fibres of the nerve roots.

8D Sensory loss (C7 dermatome) over the front and back of the index, middle, and ring fingers, and the middle part of the front and back of the hand and wrist.

8E The scapula would be 'winged' (serratus anterior); adduction would be weak (pectoralis major, latissimus dorsi); the limb would be mottled in colour because of lack of control of blood flow to the skin, and sweating reduced (autonomic nervous system).

Seminar 9

9A Muscles of the anterior compartment of the arm (biceps, brachialis, coracobrachialis) would be paralysed and flexion at the shoulder and especially the elbow would be weak. There would be sensory loss over the lateral aspect of the forearm.

9B Anaesthetic should be injected into the skin on the medial and lateral aspects of the root of the fingers (ring block) to reach the digital nerves. Catecholamines can be added to local anaesthetics to reduce blood flow and thereby localize their action; the smooth muscle of the arteries may be hypersensitive to the amines and this could lead to arterial spasm and cutting off of the blood supply to the fingers.

9C Paralysis of the muscles of the anterior compartment of the forearm with the exception of flexor carpi ulnaris: therefore flexion at the wrist, metacarpophalangeal and interphalangeal joints, and abduction at the wrist will be weak.

Seminar 10

10A Division of the ulnar nerve causes paralysis of the interossei and the medial two lumbricals, of the muscles of the hypothenar eminence, and of adductor pollicis. The hand gradually adopts a 'claw' posture with the metacarpophalangeal joints extended and the interphalangeal joints flexed (least obvious in the index and middle fingers).

Sensation would be lost over the palmar aspect of the ulnar 1½ fingers and, if the severance were proximal to the origin of the dorsal cutaneous branch, on the dorsal aspect also. Mottling over the same area of the skin, due to absence of autonomic nervous control would also be present.

10B Because the lumbricals to the middle and index fingers are usually supplied by the median nerve.

10C Contributions from flexor carpi ulnaris and flexor digitorum profundus to the little and ring fingers would also be paralysed giving weak ulnar deviation at the wrist and weak flexion of the little and ring fingers.

Seminar 11

11A Wrist drop and weakness of finger extension would be the most marked features due to paralysis of the muscles of the posterior compartment of the forearm. Triceps would probably not be impaired because its nerves arise in the axilla and upper arm. Sensory loss in the forearm might be difficult to detect because the supply of the medial and lateral cutaneous nerves of forearm wrap around onto the posterior surface, but some loss over the dorsum of the radial side of the hand should be detectable.

Seminar 12

12A If the wound affected the flexor synovial sheath, sepsis could spread proximally along the sheath to the common flexor synovial sheath at the carpal tunnel. Sepsis leads to inflammatory processes which may in turn lead to fibrous connections between muscle tendons. If such a condition occurred then movements of the fingers would be severely affected.

Seminar 13

13A Palmar skin is thick first because it is genetically programmed to be so; secondly, environmental factors usually reinforce this stimulus

13B The skin prevents microorganisms from entering the body and also, if burned badly, very considerable fluid loss can occur from the burned surface.

13C Skin creases tether the skin to the underlying palmar aponeurosis and this, with the dermal ridges, optimizes transmission of mechanical force from the musculoskeletal system to the skin in gripping.

13D The skin goes blue because the blood becomes more deoxygenated as a result of slowing of circulation in the skin (controlled by sympathetically induced vasoconstriction). Supply to deeper tissues and adequate venous return are ensured by the presence of arteriovenous anastomoses. Some of the more superficial arterio–venous anastomoses are temperature sensitive.

13E If you lick your skin you will find that it has a salty taste because eccrine sweat contains sodium chloride. Salt loss occurs as a result of heavy sweating in hot, dry atmospheres.

13F Sweat glands open out on to the surface of the ridges.

13G They enhance the sense of touch; the ridges are deformed as they pass over a textured surface (e.g. over a piece of sculpture) and this is transmitted to the underlying sensory receptors which are stimulated.

Lower limb

Seminar 1

1A The collagenous fibres on which crystals of bone salts (calcium hydroxyapatite) are deposited are laid down along lines of stresses and strains within the bone. Many of these forces result from weight bearing, though the basic trabecular architecture is genetically determined.

Seminar 2

2A The articular facets of the shoulder joint are such as would enable a wide range of movement to take place, stability in this joint depends on ligaments and, particularly, on the rotator cuff of muscles. The articular facets of the hip joint, on the other hand, provide a much greater degree of stability, commensurate with weight–bearing locomotion, and therefore a smaller degree of movement.

2B The femoral capital epiphysis lies in the upper outer quadrant, i.e. the head of the femur is dislocated; Shenton's line is discontinuous.

2C The left hip is normal since, if the line of the upper margin of the neck is extended, it passes through the upper part of the capital epiphysis. That of the right side passes well above the epiphysis.

2D It is likely that the cruciate ligaments have been ruptured.

2E It is most likely that the medial meniscus of the supporting knee will have been torn. The meniscus is held by the medial collateral ligament and is subject to rotational stress as the momentum of the body attempts to rotate the femur on the supporting tibia.

2F If the injury results from hyper–inversion of the foot then lateral structures of the ankle may be damaged—the lateral ligament (3 components) and the lateral malleolus; if the injury results from hyper–eversion then medial structures of the ankle may be damaged—i.e. the deltoid ligament and the medial malleolus.

Seminar 3

3A It is active in strong lateral rotation of the thigh; its upper fibres are active in strong abduction of the thigh. It is also a tensor of the iliotibial tract and thus may have an action on the knee joint.

Seminar 4

4A Rectus femoris helps to flex the hip joint and extends the knee; sartorius flexes and laterally rotates the hip joint and helps flex the knee.

4B Vastus medialis counteracts the pull of vastus lateralis and rectus femoris which, because the axis of the femur makes an angle with that of the tibia, tend to pull the patella laterally. Note also the larger femoral articulation for the patella on the lateral side.

4C Rectus femoris has ruptured.

4D Adductor magnus (being a hamstring in addition to an adductor) extends the femur at the hip joint; obturator externus laterally rotates the femur at the hip.

4E Run your finger upwards along the tendon of adductor longus which is easily seen and felt in the slightly abducted thigh; it is attached to the pubic bone immediately inferior to the pubic tubercle.

4F Adductor magnus has both hamstring and adductor components. Therefore it is supplied by both sciatic and obturator nerves.

Seminar 5

5A The anterior compartment muscles dorsiflex (extend) the foot at the ankle, while tibialis anterior also inverts the foot and helps to support the medial aspect of the transverse arch.

5B Peroneus longus and brevis evert the foot and plantarflex (flex) it.

5C The superficial calf muscles plantarflex the foot thereby enabling a person to stand on the ball of the foot (or the tips of the toes). Soleus is more of a 'postural' muscle while gastrocnemius wins the long jump!

5D *a*) No effect except that plantar flexion would be weaker.
   b) Inability to propel the body forwards as in walking.

Seminar 6

6A It provides a fluid but compact and compartmentalized, deformable mass whereby the weight of the body is distributed on the ground.

6B In the hand the tendons of flexor digitorum superficialis, like those of flexor digitorum brevis, split to allow those of flexor digitorum profundus to pass through to their insertion.

6C It enables Man to propel himself forward (push off) during walking or running, especially if he is using the 'balls of the feet'.

6D The medial plantar nerve is comparable in its distribution to the median nerve.

Seminar 7

7A Yes, in the form of the superficial and deep palmar arches. Note that in the foot there is no equivalent of the superficial arch, presumably due to considerations of pressure.

7B Insufficient blood supply. Determine the position of the main arteries to the leg in which pulsation can be detected and the extent of the pulsation (e.g. the tibial artery behind the medial malleolus). The fundamental cause of the trouble would have to be investigated; pharmacological inhibition or surgical interruption of the sympathetic nerve supply could be tried; vascular bypass grafts are sometimes possible. As a last resort, amputation of the devascularized area may become necessary.

7C No, since there is no tightly confined osteofascial compartment in the calf.

7D Locate the pubic tubercle (see **4E**). If the hernial swelling is above the tubercle then it is of a direct or indirect inguinal variety (having emerged through, or close to, the superficial inguinal ring); femoral hernias emerge through the femoral canal inferior and lateral to the tubercle.

Seminar 8

8A S3, 4; note that this reflects the caudal origin of this area. The lower limbs develop cranial to this region, as lateral extensions of the trunk, and subsequently rotate to their definitive position.

8B L2, 3, 4, sensory nerve roots are stimulated. Sensory (afferent) fibres synapse with motor (efferent) neurones at spinal cord levels L2, 3, 4, thus forming a monosynaptic reflex arc. The knee jerk is elicited in order to test the integrity of the sensory and motor fibres which comprise the arc and especially their excitability, which depends on influences from higher nervous centres.

Seminar 9

9A Foot drop with loss of dorsiflexion and eversion (paralysis of lateral and anterior compartment muscles); loss of sensation over the lower lateral part of the front of the leg and the dorsum of the foot.

9B Paralysis of the hamstring and calf plantar muscles (tibial nerve), in addition to the muscles of the anterior and lateral compartments of the leg (common peroneal nerve); loss of sensation below the knee. A major disability would be foot drop, which can be helped very simply by attaching a spring to the front of the shoe which is secured to a band below the knee.

Spine

1A Growth of the vertebral column during development elongates it more (in a rostrocaudal direction) than it does that of the spinal cord.

1B The side on which the leg is lifted. Erector spinae contracts to laterally flex the spine away from the supporting leg and thus restore the centre of gravity which has been disturbed.

1C At the atlanto–occipital joint, nodding movements are possible with slight lateral flexion but no rotation; at the atlanto–axial joint, the atlas rotates around the dens of the axis.
The intercervical articular facets lie on a sloping, transverse plane.
This permits flexion/extension and a small amount of lateral flexion.
The interthoracic articular facets lie on the circumference of a large circle, the centre of which is in the trunk. Thus rotation is possible here, but rather little flexion or extension, which is limited by the ribs.
The interlumbar facets also lie on the circumference of a circle but, because the centre of that circle is dorsal to the trunk, the facets form strong interlocking articulations which permit some flexion and extension but no rotation of the trunk.

1D Accessory cervical ribs give rise to more problems. T1 which forms the lower part of the brachial plexus, has to emerge from the chest and travel over the first rib to reach the plexus. If an accessory rib is present it passes over this and is stretched, especially when the arm is carrying a heavy weight.

1E The transverse ligament of the atlas prevents the odontoid peg from moving backwards and therefore damaging the medulla and upper cervical spinal cord.

1F The alar ligaments restrain both rotation and flexion/extension between the occiput and the axis.

1G To open up the area between the 4th and 5th lumbar spines where the lumbar puncture needle is being inserted.

1H No, the extradural space contains fat and the internal vertebral venous plexus.

1I Paralysis and loss of sensation below the level of the lesion. Autonomic responses would lack higher control, therefore loss of coordination of pelvic visceral reflexes.
The patient would have the use of the diaphragm (C3, 4, 5) and could therefore breathe.
C5, 6 remain intact therefore most shoulder joint movements would be possible, though weaker; abduction of the shoulder being most affected. At the elbow, flexion would be possible and weak flexion of the wrist might be possible.

1J Diaphragmatic and intercostal breathing would be intact, as would sympathetic reflexes.

1K This lesion, which is relatively common, could give rise to pain down the lower and lateral aspect of the leg, passing into the foot (S5 dermatome). It would also lead to weakness of eversion of the foot.

Appendix

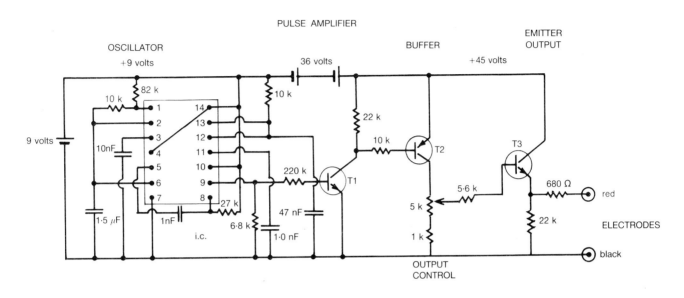

Fig. A.1 **Circuit diagram of an electronic stimulator for testing muscles. (T1: ZTX304; T2: ZTX504; T3:ZTX304; i.c.: NE556A.)**

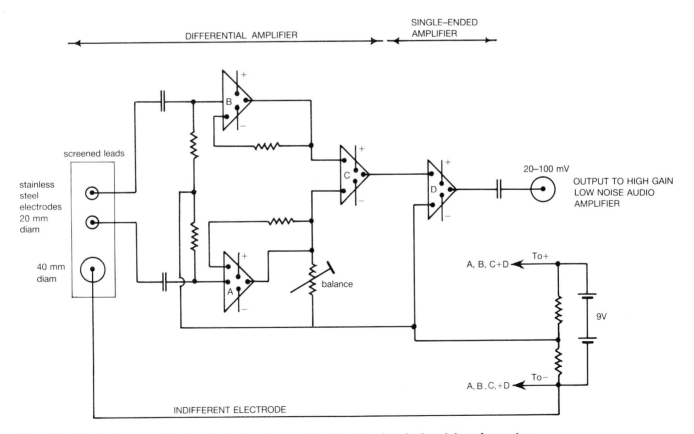

Fig. A.2 **Block diagram of a simple electronic amplifier for the electrical activity of muscles.**

Index